CM
Swan

METHODS TO STUDY LITTER DECOMPOSITION

Methods to Study Litter Decomposition

A Practical Guide

Edited by

MANUEL A.S. GRAÇA
University of Coimbra, Portugal

FELIX BÄRLOCHER
Mount Allison University, Canada

and

MARK O. GESSNER
Limnological Research Center, Switzerland

 Springer

A C.I.P. Catalogue record for this book is available from the Library of Congress.

ISBN-10 1-4020-3348-6 (HB) Springer Dordrecht, Berlin, Heidelberg, New York
ISBN-10 1-4020-3466-0 (e-book) Springer Dordrecht, Berlin, Heidelberg, New York
ISBN-13 978-1-4020-3348-3 (HB) Springer Dordrecht, Berlin, Heidelberg, New York
ISBN-13 978-1-4020-3466-4 (e-book) Springer Dordrecht, Berlin, Heidelberg, New York

Published by Springer,
P.O. Box 17, 3300 AA Dordrecht, The Netherlands.

Printed on acid-free paper

TABLE OF CONTENTS

PART 3. MICROBIAL DECOMPOSERS

PART 4. ENZYMATIC CAPABILITIES

PART 5. DETRITIVOROUS CONSUMERS

PART 6. DATA ANALYSIS

PREFACE

Forests are the most common terrestrial biomes, covering about 30% of the earth's land surface. Generally, about 10% of the primary production in forests is used by herbivores, while the remaining 90% enter the decomposition loop as plant litter, which includes leaves, wood, roots and other plant parts. Leaf litter production in forests ranges from about 100 to 1400 g dry mass m^{-2} $year^{-1}$, suggesting that decomposition is a vital process in the functioning of forest soils and forest ecosystems as a whole. Similar cases can be made for low-order woodland streams, where plant litter is a major energy source, for grasslands and wetlands, and for a range of other both aquatic and terrestrial ecosystems.

Decomposition of plant matter is a complex process that involves bacteria, fungi and invertebrates as well as physical-chemical processes. It is influenced by the chemical and physical properties of the decomposing plant material and by environmental conditions such as temperature, moisture and nutrient availability. A thorough understanding of these processes is critical not just to grasp the essence of ecosystem functioning but to predict consequences of global environmental change on carbon budgets at various scales. Because decomposition greatly affects the balance between carbon dioxide returned to the atmosphere and long-term carbon sequestration, solid knowledge of the rates, pathways and controls of decomposition has become imperative to making informed ecological forecasts of carbon cycling in altered environmental conditions and of the feedbacks on climate.

Given the significance of the process, it is not surprising that ecologists have studied litter decomposition at least since Darwin. An upsurge occurred when the use of litter confined in boxes and open-ended tubes, later in mesh bags, was introduced in the 1930s. Methods have since been substantially broadened and refined, even though some basic approaches such as the mesh-bag technique are still useful and widely employed.

This book is the outcome of several editions of an international "Advanced Course on Litter Decomposition" held since 1998 at the University of Coimbra, Portugal. The courses aimed at introducing graduate students and postdoctoral scientists to a range of methods that have been used to advance understanding of the litter decomposition process. Emphasis was on ecological field methods, associated chemical, microbiological and enzymatic analyses, and involvement of detritivorous invertebrates. The idea of publishing the experimental protocols presented in the courses followed the request from participants and a number of colleagues wishing to have the teaching material in printed form. Consequently, the focus of this "recipe" book is on the description of specific procedures. A companion book addressing the background of different methods and evaluating the usefulness in different contexts is currently in preparation.

The primary target audience of this book includes graduate students, instructors of specialized and general ecology courses, and professional scientists working on organic matter dynamics in both aquatic and terrestrial ecosystems. Most of the contributing authors have a research background in stream ecology. This bias is reflected in the book. Some of the described methods will thus be specific to streams and rivers; others will require adaptation when applied to other ecosystems. However, the majority of proposed procedures will be applicable to aquatic and terrestrial systems alike, and many will also be useful to characterize fractions of organic matter – and processes and biota associated – other than decomposing plant litter in natural environments.

A number of individuals have contributed to this book. We are particularly grateful to the participants of the courses for their questions and comments which improved the presented protocols. Our special thanks go, of course, to the authors of the chapters for delivering their manuscripts on time (mostly) and particularly for their great patience during the editing phase when confronted with our numerous queries, comments and suggestions. We believe this effort was worthwhile and hope users of the methods will share this perception.

Coimbra, Sackville and Kastanienbaum,

Manuel A.S. Graça
Felix Bärlocher
Mark O. Gessner

PART 1

LITTER DYNAMICS

CHAPTER 1

LITTER INPUT

ARTURO ELOSEGI & JESÚS POZO

Departamento de Biología Vegetal y Ecología, Facultad de Ciencia y Tecnología
Universidad del País Vasco, C.P. 644, 48080 Bilbao, Spain.

1. INTRODUCTION

Allochthonous plant litter is a major source of carbon and energy for stream organisms, especially in narrow reaches where riparian cover limits primary production (Vannote et al. 1980, Webster & Meyer 1997). Typically, several hundred grams of litter dry mass per square meter of stream bed are received per year (Table 1.1). Even when macrophytes are abundant, the detrital pathway driven by allochthonous inputs is important for stream communities (Hill & Webster 1983).

Litter inputs can be quantified for a whole stream or selected reaches (Cummins et al. 1983, Minshall 1996). It is important to realize the difference, because upstream import to single reaches can be substantial, but plays a small role when entire stream systems are considered. However, where reaches along a stream differ greatly in riparian vegetation or bank characteristics, whole-stream studies require extensive sampling to be meaningful.

A large part of the litter entering a stream channel consists of leaves from riparian vegetation, particularly in forested streams. Inputs include transport from upstream, direct (or vertical) inputs (also called fall-in), and lateral inputs of material deposited on the forest floor and mobilized by wind or some other agent (also called blow-in). Another significant contribution is made by woody debris (Díez et al. 2001). A large part of the wood inputs occur because of debris torrents, landslides, unusual storms and similar events (Harmon et al. 1986). Because of their sporadic occurrence, these inputs are not easily measured by routine procedures. Consequently, it is often useful to differentiate between inputs of coarse wood and other sorts of litter.

Of all pathways of litter input, transport from upstream is by far the most difficult to measure. It can be highly variable in response to discharge, and is often impossible to estimate accurately during high flow. A large portion of litter deposited in the stream channel and on the stream banks is mobilized during spates (Webster et al. 1990), which tend to be unpredictable and hence difficult to sample.

3

M.A.S. Graça, F. Bärlocher & M.O. Gessner (eds.), Methods to Study Litter Decomposition:
A Practical Guide, 3 – 12.

Therefore, measurements of long-term litter transport in streams tend to be gross underestimates, even when they are based on frequent sampling schedules (Golladay 1997). This shortcoming can be partly corrected for by plotting stream discharge versus the concentration of drifting litter from a long time-series and extrapolating to the concentrations expected during floods (Webster et al. 1990). Nevertheless, discharge-concentration curves typically yield poor fits (Gurtz et al. 1980). The first flood after a long autumn base flow period will scour a much larger amount of litter than similar floods later in the season. Furthermore, the relation between concentration and discharge is characterized by hysteresis, with litter concentrations being typically lower during the falling limb of a storm hydrograph (Williams 1989), because most of the litter deposited near the stream channel has already been scoured away, thus limiting further mobilization (Williams 1989). For short reaches, one may assume that drift inputs equal drift outputs, so that only vertical and lateral inputs need to be considered.

Table 1.1. Vertical and lateral litter inputs to selected streams.

Location	Vegetation	Stream order	Vertical input $(g\ m^{-2}\ y^{-1})$	Lateral input $(g\ m^{-2}\ y^{-1})$	Reference
Alaska, USA	Tundra	4	0	500	1
Québec, Canada	Taiga	1	417	344	2
Québec, Canada	Taiga	2	217	56	2
Oregon, USA	Coniferous forest	1	537	667	3
Germany	Temperate mixed forest	1	700	–	3
Spain	Temperate deciduous forest	1	611	104	4
Spain	Eucalyptus plantation	1	478	24	4
Portugal	Temperate deciduous forests	1—2	261	–	5
Portugal	Eucalyptus plantations	1—2	204	–	5
Arizona, USA	Desert shrubs	5	17	3	6
Georgia, USA	Temperate mixed forest	6	843	3520	7
Brazil	Tropical gallery forest	3	713	421	8
Australia	Temperate eucalyptus forest	3	617	61	9

1 = Harvey et al. (1997); 2 = Naiman & Link (1997); 3 = Benfield (1997); 4 = Pozo et al. (1997); 5 = Abelho & Graça (1996); 6 = Jones et al. (1997); 7 = Meyer et al. (1997); 8 = Afonso et al. (2000); 9 = Campbell et al. (1992).

In several studies, ways have been sought to sample continuously a portion of the flowing water, or to build grid-like structures to retain all litter transported during extended periods (Likens & Borman 1995). However, these approaches require large effort and are seldom feasible, especially when streams are larger than first order. Here, we describe a less accurate but more readily applicable method for nonwoody litter inputs to small streams. Vertical inputs are collected with litter-fall baskets, blow-in inputs with lateral traps, and transport inputs with drift nets (Webster & Meyer 1997).

2. SITE SELECTION AND EQUIPMENT

2.1. Site Selection

Litter inputs can be measured in any stream but they are most meaningful in small streams surrounded by riparian forests. Although logistical difficulties, measurement uncertainties and hazards are normally smaller in narrow, shallow streams, even large rivers can be studied. A springbrook may be selected to avoid measuring inputs from upstream. Reach length depends on the objectives of the research and on the variability of the riparian areas. A 100-m reach with no tributaries will suffice for most purposes. Streams with extensive floodplains are more difficult to study, as most inputs are likely to occur during floods.

2.2. Equipment and Material

- Litter-fall traps (Fig. 1.1) can be constructed from plastic laundry baskets, or by sewing 1-mm mesh to any wooden or metallic frame. Although 0.25-m^2 traps appear to be most popular, traps ranging from 0.025 to 1 m^2 are described in the literature. More replicates are necessary when small traps are used. Traps must allow rainwater to drain quickly, but ensure that no material larger than 1 mm is lost. The number of traps necessary for reliable estimates depends on spatial variations of the riparian forest, but for forest stream reaches with full canopies, ten are enough in most cases.

Figure 1.1. A typical trap for determining vertical litter inputs. The mesh is fixed to a wooden or metallic frame hanging from nearby trees by four ropes.

- Lateral-input traps (Fig. 1.2) can be constructed by tying a 1-mm mesh to a rectangular wooden or metallic frame. The ideal trap size is about 50 cm wide and 20 cm high. As with the traps for measuring vertical inputs, 10 lateral traps are enough for most purposes.

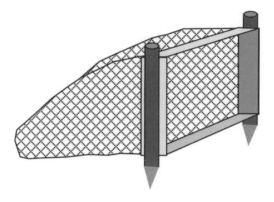

Figure 1.2. A typical trap for determining lateral or blow-in litter inputs. The mesh is fixed to a wooden or metallic frame and secured to two stakes.

- Drift nets (Fig. 1.3) typically have a rectangular or square mouth and a long funnel (1-mm mesh size) to delay clogging. Additional features can include rings in the frame to fix the net to the stream bottom with stakes, and a plastic tube fixed at the end of the net to collect CPOM retained. To sample during high flow, 3—5 nets should be used in all but the narrowest reaches.

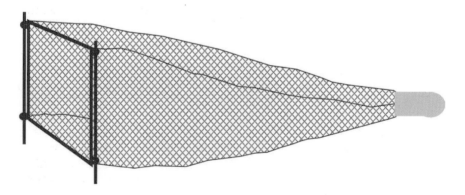

Figure 1.3. A typical drift net to sample CPOM in transport. The mesh is fixed to a metallic frame. Two metal rods can be used to secure the net in soft-bedded streams. A tube fixed at the end of the net makes sample collection easier.

- Aluminium trays
- Balance
- Crucibles
- Current meter
- Desiccator
- Drying oven
- Freezer
- Hammer
- Large zip-lock plastic bags
- Measuring stick and ruler
- Measuring tape
- Muffle furnace
- Plastic trays
- Random number table
- Ropes
- Tongs

3. EXPERIMENTAL PROCEDURES

3.1. Reach Preparation

1. Suspend litter-fall baskets over the stream channel by tying them to nearby trees with ropes. Make sure the traps will not become submerged during the highest expected flood, but place them as low as possible above the water level. Alternatively, in narrow streams under fully closed canopies, baskets can be placed on the banks, fixed to stakes. Distribute baskets randomly. Extend a measuring tape along the study reach. Take a number from a random-number table and go to the corresponding tape number. Extend another measuring tape across the stream, and select the basket location with a second number. Repeat for each basket. Exposing baskets in 'typical' places is not random sampling. Tag each basket with a number.
2. Place lateral-input traps randomly on the bank, perpendicularly to the stream channel. Position the frame vertically to avoid direct inputs, and fix it tightly to stakes. The frame must be at ground level to allow free entry of CPOM on the forest floor. Tag each trap with a number.
3. Measure the surface of the reach. Take 10—20 regularly spaced measurements of channel width, average them, and multiply by channel length. For more accurate estimates, prepare a detailed map of the reach, and measure surface area from the map. The wetted channel can vary greatly with discharge. Therefore, map the whole channel.

3.2. Sampling

1. Every other week, or at least every month, collect the material in litter-fall baskets and lateral traps, and enclose it in plastic bags. Mark each bag with the basket or trap number and the date. Discard any branches larger than 1 cm in diameter.
2. With at least the same frequency, sample inputs from upstream. If possible, locate a drift net in a narrow place where all stream water is funnelled into the net, in the upper end of the study reach. Keep the net in place for 4 h, or as long as the material retained is not clogging the net. In the latter case, measure the time the net has been screening water. When the stream is too wide to be funneled in a single net, preferably distribute several nets across the channel or use a stop net covering the entire stream width. Measure the cross-sectional area intercepted by each net and the water velocity at their mouths, as soon as possible after setting the nets. Measure again the cross-sectional area and velocity of the intercepted water just before removing the nets. Enclose the collected material individually in plastic bags. Mark each bag.
3. Measure the cross-sectional area of the stream, and the water velocity, to calculate stream discharge.
4. Preferably repeat transport measurements several times during a storm.
5. Carry the collected material to the laboratory. If it can be processed within the next few days, leave it air-drying in open plastic trays. Mark each tray with the sample identification. Otherwise, freeze material as soon as possible.

3.3. Laboratory Procedures

1. Sort samples into leaves, fruits, bark, twigs and other materials. Sort leaves by species. Put each of these categories in an aluminium tray. Mark all trays with the material and sample identifications.
2. Dry all samples at 40—50 °C to constant mass.
3. Cool the trays in a desiccator and weigh them.
4. Transfer the material to pre-weighed crucibles.
5. Ash crucibles at 500 °C for 4 h.
6. Cool the crucibles in a desiccator and weigh them.

3.4. Calculations

1. Calculate ash-free dry mass (AFDM) of each category by subtracting the ash mass from that of the dry matter.
2. Sum all categories in a sample to get the total AFDM and divide by the basket surface. Express results in g AFDM m^{-2} or in g AFDM m^{-2} d^{-1}.
3. To calculate the total amount of vertical inputs between sampling periods, multiply the above figure by the surface area of the reach.

4. Divide AFDM of lateral inputs by the trap length. Express results in g AFDM m^{-1} or in g AFDM m^{-1} d^{-1}.
5. To calculate the total amount of lateral inputs, multiply the above figure by the measured bank length. To calculate lateral inputs on an area basis, divide by the surface area of the reach.
6. To calculate the volume of water filtered by each drift net, multiply the cross-sectional area of the water funneled into the net by the average water velocity measured immediately both after introducing and before removing the net.
7. To calculate the concentration of CPOM transported from upstream, divide the AFDM of transport inputs by the filtered water volume. Express results in g AFDM m^{-3}. When more than one net has been used simultaneously, calculate the average concentration of CPOM in water during this period.
8. To calculate CPOM inputs from upstream, multiply the above concentration by stream discharge and by the time elapsed between samplings. It is worth exploring discharge – concentration relations. If the regression is significant and continuous discharge data are available, calculate CPOM inputs from this regression.
9. Divide CPOM inputs by the surface area of the reach to calculate the per-metre contribution of transport from upstream to total inputs.

4. FINAL REMARKS

A relatively large number of replicates, both in time and space, are necessary to get reliable data, and the exact location of traps and nets can significantly affect conclusions.

To be ecologically meaningful, litter collections should at least encompass the main period of leaf fall (i.e., autumn under deciduous temperate forests), and preferably a whole year. Even then, a great deal of caution is necessary when making long-term extrapolations, as inputs are far from constant from year to year (Cummins et al. 1983). Ideally, the measurements should begin before the onset of the main leaf-fall period, especially if one is interested in annual data. A small advance of the leaf fall due to unusual weather can strongly affect calculations.

5. REFERENCES

Abelho, M. & Graça, M.A.S. (1996). Effects of eucalyptus afforestation on leaf litter dynamics and macroinvertebrate community structure of streams in central Portugal. *Hydrobiologia*, 324, 195-204.

Afonso, A.A.O., Henry, R. & Rodella, R.C.S.M. (2000). Allochthonous matter input in two different stretches of a headstream (Itatinga, São Paulo, Brazil). *British Archives of Biology and Technology*, 43, 335-343.

Benfield, E.F. (1997). Comparison of litterfall input to streams. In: J.R. Webster & J.L Meyer (eds.), Stream organic matter budgets. *Journal of the North American Benthological Society*, 16, 3-161. pp. 104-108.

Campbell, I.C., James, K.R., Hart, B.T. & Devereaux, A. (1992). Allochthonous coarse particulate organic material in forest and pasture reaches of two south-eastern Australian streams. I. Litter accession. *Freshwater Biology*, 27, 341-352.

Cummins, K.W., Sedell, J.R., Swanson, F.J., Minshall, G.W., Fisher, S.G., Cushing, C.E., Petersen, R.C. & Vannote, R.L. (1983). Organic matter budgets for stream ecosystems: problems in their evaluation. In: J.R. Barnes & G.W. Minshall (eds.). *Stream ecology. Applications and testing of general ecological theory.* (pp. 299-353). Plenum Press, New York.

Díez, J.R., Elosegi, A. & Pozo, J. (2001). Woody debris in north Iberian streams: influence of geomorphology, vegetation and management. *Environmental Management,* 28, 687-698.

Golladay, S.W. (1997). Suspended particulate organic matter concentration and export in streams. In: J.R. Webster & J.L. Meyer (eds.), Stream organic matter budgets. *Journal of the North American Benthological Society*, 16, 3-161, pp. 122-131.

Gurtz, M.E., Webster, J.R. & Wallace, J.B. (1980). Seston dynamics in southern Appalachian streams: effects of clear-cutting. *Canadian Journal of Fisheries and Aquatic Sciences*, 37, 624-631.

Harmon, M.E., Franklin, J.F., Swanson, F.J. Sollins, P., Gregory, S.V., Lattin, J.D., Anderson, N.H., Cline, S.P., Aumen, N.G., Sedell, J.R., Lienkaemper, G.W., Cromack, K. & Cummins, K.W. (1986). Ecology of coarse woody debris in temperate ecosystems. *Advances in Ecological Research*, 15, 133-302.

Harvey, C.J., Peterson, B.J., Bowden, W.B., Deegan, L.A., Finlay, J.C., Hershey, A.E. & Miller, M.C. (1997). Organic matter dynamics in the Kuparuk River, a tundra river in Alaska, USA. In: J.R. Webster & J.L. Meyer (eds.), Stream organic matter budgets. *Journal of the North American Benthological Society*, 16, 3-161, pp. 18-23.

Hill, B.H. & Webster, J.R. (1983). Aquatic macrophyte contribution to the New River organic matter budget. In: Fontaine, T. & Bartell, S. (eds.), *Dynamics of Lotic Ecosystems* (pp. 273-282). Ann Arbor Science. Michigan.

Jones, J.B., Schade, J.D., Fisher, S.G. & Grimm, N.B. (1997). Organic matter dynamics in Sycamore Creek, a desert stream in Arizona, USA. In: J.R. Webster & J.L. Meyer (eds.), Stream organic matter budgets. *Journal of the North American Benthological Society* 16, 3-161, pp. 78-82.

Lienkaemper, G.W. & Swanson, F.J. (1987). Dynamics of large woody debris in streams in old-growth Douglas-fir forests. *Canadian Journal of Forest Research,* 17,150-156

Likens, G.E. & Bormann, F.H. (1995). *Biogeochemistry of a Forested Ecosystem.* 2nd edition. Springer Verlag. New York.

Meyer, J.L., Benke, A.C., Edwards, R.T. & Wallace, J.B. (1997). Organic matter dynamics in the Ogeechee River, a blackwater river in Georgia, USA. In: J.R. Webster & J.L. Meyer (eds.), Stream organic matter budgets. *Journal of the North American Benthological Society*, 16, 3-161, pp. 82-87.

Minshall, G.W. (1996). Organic matter budgets. In: Hauer, F.R. & Lamberti, G.A. (eds.), *Methods in Stream Ecology* (pp. 591-605). Academic Press. San Diego.

Naiman, R.J. & Link, G.L. (1997). Organic matter dynamics in 5 subarctic streams, Quebec, Canada. In: J.R. Webster & J.L Meyer (eds.), Stream organic matter budgets. *Journal of the North American Benthological Society*, 16, 3-161, pp. 33-39.

Pozo, J., González, E., Díez, J.R., Molinero, J. & Elósegui, A. (1997). Inputs of particulate organic matter to streams with different riparian vegetation. *Journal of the North American Benthological Society*, 16, 602-611.

Vannote, R.L., Minshall, G.W., Cummins, K.W., Sedell, J.R. & Cushing, C.E. (1980). The river continuum concept. *Canadian Journal of Fisheries and Aquatic Sciences*, 37, 130-137.

Webster, J.R. & Meyer, J.L. (eds.) (1997). Stream organic matter budgets. *Journal of the North American Benthological Society*, 16, 3-161.

Webster, J.R., Golladay, S.W., Benfield, E.F., D'Angelo, D.J. & Peters, G.T. (1990). Effects of forest disturbance on particulate organic matter budgets of small streams. *Journal of the North American Benthological Society*, 9, 120-140.

Williams, G.P. (1989). Sediment concentration versus water discharge during single hydrologic events in rivers. *Journal of Hydrology*, 111, 89-106.

CHAPTER 2

LEAF RETENTION

ARTURO ELOSEGI

*Departamento de Biología Vegetal y Ecología, Facultad de Ciencia y Tecnología
Universidad del País Vasco, C.P. 644, 48080 Bilbao, Spain.*

1. INTRODUCTION

Allochthonous organic matter, especially leaf litter, is the main energy source of food webs in headwater forested streams (Vannote et al. 1980, Lohman et al. 1992, Cummins et al. 1989). Leaf litter enters streams mainly in a large burst during the period of leaf abscission (autumn in most temperate areas), and can be either trapped in the reach and thus become available for heterotrophs, or transported downstream. Therefore, the capacity of a stream reach to retain materials (retentiveness) is important for the productivity and ecosystem efficiency of streams (Bilby & Likens 1980, Pozo et al. 1997).

Channel morphology is a key factor determining the capacity of streams to retain leaf litter, with narrow, rough-bottom streams being most retentive (Webster et al. 1994, Mathooko et al. 2001). Wood, especially when forming debris dams, enhances leaf retention and storage (Bilby & Likens 1980, Raikow et al. 1995, Díez et al. 2000). Changes in stream stage produce temporal variations in retention capacity, as higher discharge results in larger depth and width, and higher hydraulic power, thus decreasing retentiveness (Ehrman & Lamberti 1992, Larrañaga et al. 2003).

Measuring leaf litter retention over short periods involves releasing leaves and estimating their downstream displacement. Four types of "leaves" may be used: (1) Natural leaves with some kind of mark that does not modify their short-term behaviour in water. Many paints affect leaf buoyancy and stiffness, so care must be taken to select a good dye; alternatively, a narrow line can be painted on both sides. The colours most easily recognized in streams are bright blue and blaze orange. (2) Leaves that do not occur naturally in the stream can sometimes be easily recognized. The bright yellow leaves of the exotic ginkgo tree (*Ginkgo biloba*), collected in autumn and stored dry between paper sheets, have often been used. (3) Artificial leaves of ornamental plastic plants, which can be painted in easily recognized

colours. Although such artificial leaves may closely resemble natural ones, their floating behaviour can be different. (4) Any other material that is easily seen and

*M.A.S. Graça, F. Bärlocher & M.O. Gessner (eds.), Methods to Study Litter Decomposition:
A Practical Guide*, 13 – 18.

behaves like leaves. Most commonly, strips (ca. 3 × 10 cm) from different types of plastic and in different colours are prepared.

Artificial materials are cheap and easily available throughout the year, whereas the use of natural leaves requires advanced planning, rather time-consuming collection, drying and storage. Furthermore, natural leaves tend to fragment easily, and may therefore not be used repeatedly. However, released "leaf" material can be lost in the study reaches, especially when single-point collections are made (see below). Artificial or painted leaves should therefore only be used if one is confident that all leaves will be recovered. If this is not the case, it is preferable to use natural exotic leaves.

It is important to check that the materials used behave like natural leaves in a stream (see Fig. 2.1). If the goal is simply to compare the retention capacity of different reaches, any material can be used, but if the goal is to simulate the retention of real leaves, different kinds of materials need to be calibrated against the riparian leaf species most abundant in the study area. This is a point worth exploring in detail, as differences between materials can be substantial (Fig. 2.1; Young et al. 1978, Prochazka et al. 1991, Canhoto & Graça 1998), and the relationship between retention and leaf morphology is not straightforward (Larrañaga et al. 2003). Thus, a great deal of caution is necessary when comparing streams based on results obtained with different materials.

This chapter presents a method to measure the capacity of stream reaches to retain leaf litter in the short term. This is done by monitoring the downstream displacement of leaves released at one point. The average travel distance of the leaves is calculated by plotting the proportion of leaves in transport at a given point against the measured travel distance and fitting the data to an exponential decay model. This approach assumes that the number of leaves retained at any one point along the experimental reach is directly proportional to the number of leaves in transport. Two methods are given here:

The multiple-point collection method is best suited for small clear-water streams. All leaves are retrieved, and the distance travelled by each leaf is measured. A net is placed downstream of the reach as a safety device, to prevent the loss of leaves drifting past this point.

The single-point collection method may be slightly less accurate but is useful in larger reaches or turbid waters where many leaves are unlikely to be recovered after release. Instead of measuring the distance travelled by individual leaves, the proportion of leaves reaching a net placed downstream of an experimental reach is measured. If differentially marked leaves are released at various distances from the net, an exponential regression can be calculated as with the multiple-point collection method. Unlike in multiple-point collections, the distance between release point and net is critical. The experiment is not valid if more than 90% or less than 10% of leaves reach the net (Lamberti & Gregory 1996). Whereas both natural and artificial leaves can be used with multiple-point collection, only natural leaves should be used with single-point collection, to avoid polluting the reach.

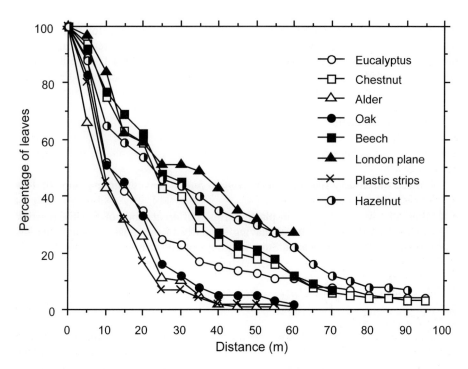

*Figure 2.1. Downstream decrease in the number of leaves transported in a third-order stream, expressed as a percentage of leaves released. Note that alder and plastic strips (3 × 10 cm) were most readily retained (average travel distance = 11.2 m), whereas London plane (*Platanus × acerifolia*) leaves travelled furthest (average travel distance = 50 m). Data from Larrañaga et al. (2003).*

Because retention distance varies with stream stage, inter-stream comparisons should be performed under similar hydrological conditions. This is most easily done during base-flow conditions, but distances so measured can grossly underestimate average retention efficiency, as leaves are more easily scoured during high flow. The relationship between travel distance and discharge can be studied for each reach by repeating retention experiments under different discharge conditions. In this case, results from different reaches can be compared even if they do not correspond exactly to the same hydrological condition.

2. SITE SELECTION AND EQUIPMENT

2.1. Site Selection

Leaf retention can be measured in almost any stream or river, but measurements are easier in wadeable streams with clear water. In first- to second-order streams most leaves are retained within a few metres during base-flow, but retention distance can increase to some tens of metres at higher discharge. Appropriate reach lengths are therefore normally 10—50 m. In larger streams and rivers, reaches 100—500 m long are recommended. As a rule of thumb, reach length should be 10 times the wetted channel width (Lamberti & Gregory 1996), but especially with the single-point collection method, it is worth running preliminary experiments to determine the most appropriate length.

2.2. Equipment and Material

- Collect and air-dry recently fallen *Gingko biloba* leaves. Store groups of 100 leaves.
- Alternatively, use 3 × 10 cm plastic strips. Different kinds of flexible plastics can be used. Select the one behaving similarly to the most abundant riparian leaf species in the study area. Store in groups of 100 strips.
- Stop net (1—2 cm mesh-size) per reach, wider than the stream channel
- Measuring tape
- Rope to tie the stop net to trees or other features
- Current meter
- Measuring stick and ruler

3. EXPERIMENTAL PROCEDURES

3.1. Field Procedures

1. The day before the experiment, soak the leaves overnight in water to give them neutral buoyancy.
2. Block the downstream end of the reach with the stop net.
3. Standing in the upstream end of the reach, release leaves or plastic strips one by one into the water. One hundred leaves are normally enough for first-order reaches, 500 for third-order reaches. Larger numbers are necessary with the single-point collection method.
4. Allow the stream to disperse leaves for one hour.

3.1.1. Multiple-Point Collection:
1. Extend measuring tape along the reach.
2. One hour after release, recover leaves that reached the stop net. Keep the net in place. Record the number of leaves.
3. Walking upstream from the net, recover all leaves. Record to the nearest metre (5 m in reaches >100 m) the distance travelled by each leaf.
4. For more exhaustive analysis, record the structure retaining each leaf (e.g. pool, riffle, channel margin, wood piece, roots, debris dam, boulder, gravel, sand).
5. After recovering all leaves, remove the stop net.

3.1.2. Single-Point Collection:
1. One hour after release, recover and count the leaves that reached the net.
2. With either method, measure stream discharge after the retention experiment.
3. Additional information of interest may be channel gradient, average width and depth, bank slope, area covered by riffles and pools, or area covered by different substrate categories (sand, gravel etc.). Of particular significance is the abundance of woody debris, as it is one of the most retentive structures found in stream channels.

3.2. Calculations

1. The number of released leaves in transport is plotted against travel distance and the data are fitted to the exponential decay model (Young et al. 1978):

$$L_d = L_0 \cdot e^{-k \cdot d} \tag{2.1}$$

2. In the single-point collection method, L_d is the number of leaves recovered at the net, L_0 is the number of leaves released, d is the distance in metres between the release-point and net, and k is the instantaneous retention rate, which is independent of reach length and number of released strips.
3. In the multiple-point collection, L_0 is the total number of leaves recovered (which should be close to the number released), and L_d the number of leaves still in transport at distance d. This is calculated by subtracting the number of leaves retained between release point and distance d from the total number of leaves recovered in the experiment.
4. Calculations can be made with any standard statistical software or a calculator. Exponential regressions are calculated by first linearizing the data by applying the natural logarithm, and then calculating the linear regression. Alternatively, and more accurately, non-linear curve-fitting may be used (see Chapter 6). The slope of the regression is the instantaneous retention rate.
5. Calculate the average travel distance as $1/k$ (Newbold et al. 1981).
6. Analysis of covariance (ANCOVA) or other approaches (see Chapter 6) can be used to test for statistically significant differences between slopes. When the multiple-point collection is chosen, the percentage of leaves retained by different channel structures can also be calculated.
7. Additionally, the relative retention efficiency of each substrate structure can be determined. To do this, simply divide the percentage of strips retained by a given structure by the percentage of wetted streambed area covered by the same structure.

4. REFERENCES

Bilby, R.E. & Likens, G.E. (1980). Importance of organic debris dams in the structure and function of stream ecosystems. *Ecology*, 61, 1107-1113.

Canhoto, C. & Graça, M.A.S. (1998). Leaf retention: a comparative study between two stream categories and leaf types.*Verhandlungen der Internationalen Vereinigung für Limnologie*, 26, 990-993.

Cummins, K.W., Wilzbach, M.A., Gates, D.M., Perry, J.B. & Taliaferro, W.B. (1989). Shredders and riparian vegetation. *BioScience*, 39, 24-30.

Díez, J.R., Larrañaga, S., Elosegi, A. & Pozo, J. (2000). Effect of removal of wood on streambed stability and retention of organic matter. *Journal of the North American Benthological Society*, 19, 621-632

Ehrman, T.P. & Lamberti, G.A. (1992). Hydraulic and particulate matter retention in a 3rd-order Indiana stream. *Journal of the North American Benthological Society*, 11, 341-349.

Lamberti, G.A. & Gregory, S.V. (1996). Transport and retention of CPOM. In: F.R. Hauer & G.A. Lamberti (eds.), *Methods in Stream Ecology* (pp. 217-229). Academic Press. San Diego.

Larrañaga, S., Díez, J.R., Elosegi, A. & Pozo, J. (2003). Leaf retention in streams of the Agüera basin (northern Spain). *Aquatic Sciences*, 65: 158-166.

Lohman, K., Jones, J.R. & Perkins, B.D. (1992). Effects of nutrient enrichment and flood frequency on periphyton biomass in northern Ozark streams. *Canadian Journal of Fisheries and Aquatic Sciences*, 49, 1198-1205.

Mathooko, J.M., Morara, G.O. & Leichtfried, M. (2001). Leaf litter transport and retention in a tropical Rift Valley stream: an experimental approach. *Hydrobiologia*, 443, 9-18.

Newbold, J.D., Elwood, J.W., O'Neill, R.V. & VanWinkle, W. (1981). Measuring nutrient spiralling in streams. *Canadian Journal of Fisheries and Aquatic Sciences*, 38, 860-863.

Pozo, J., González, E., Díez, J.R. & Elosegi, A. (1997). Leaf litter budgets in two contrasting forested streams. *Limnética*, 13, 77-84.

Prochazka, K., Stewart B.A. & Davies, B.R. (1991). Leaf litter retention and its implications for shredder distribution in two headwater streams. *Archiv für Hydrobiologie*, 120, 315-325.

Raikow, D.F., Grubbs, S.A. & Cummins, K.W. (1995). Debris dam dynamics and coarse particulate organic matter retention in an Appalachian Mountain stream. *Journal of the North American Benthological Society*, 14, 535-546.

Vannote, R.L., Minshall, G.W., Cummins, K.W., Sedell J.R. & Cushing, C.E. (1980). The river continuum concept. *Canadian Journal of Fisheries and Aquatic Sciences*, 37, 130-137.

Webster, J.R., Covich, A.P., Tank, J.L. & Crockett, T.V. (1994). Retention of coarse organic particles in the southern Appalachian Mountains. *Journal of the North American Benthological Society*, 13, 140-150.

Young, S.A., Kovalak, W.P. & Del Signore, K.A. (1978). Distances travelled by autumn leaves introduced into a woodland stream. *American Midland Naturalist*, 100, 217-222.

CHAPTER 3

MANIPULATION OF STREAM RETENTIVENESS

MICHAEL DOBSON

Department of Environmental & Geographical Sciences, Manchester Metropolitan University, Chester Street, Manchester, M1 5GD, United Kingdom.

1. INTRODUCTION

Leaf litter is the dominant energy resource in low-order shaded streams, but is only available to the vast majority of detritivores and microbial decomposers when retained on the streambed. Therefore, the retention capacity of the channel is crucial in determining the overall decomposition of litter in a stream reach or entire stream.

Occasionally one may wish to manipulate the retentive capacity of a stream channel, in order to test hypotheses about the role of physical channel attributes and related parameters in leaf litter dynamics, including litter decomposition. Manipulation of channel retentiveness can be in the form of enhancement or reduction. The actual procedures for these manipulations are straightforward, but several different techniques may be employed.

The method presented here for enhancing litter retention has been adapted from Dobson et al. (1995). It consists of deploying on the stream bed a set of litter traps each made of two steel poles connected by a piece of plastic mesh screen. The method was originally designed to investigate the influence of increased retentiveness in discrete patches on localized and overall numbers and biomass of detritivores and coarse benthic organic matter in low-order streams.

2. EQUIPMENT AND MATERIAL

2.1. Equipment

- Surber-type sampler
- Drying oven (40—50 °C)
- Muffle furnace
- Top-loading balance

M.A.S. Graça, F. Bärlocher & M.O. Gessner (eds.), Methods to Study Litter Decomposition:
A Practical Guide, 19 – 24.

2.2. Materials

- Rigid plastic mesh (mesh size ca. 8—10 mm is normally adequate), cut into 20 × 15 cm rectangles
- Steel poles (e.g. rebars). These must be narrower in diameter than the mesh used. A range of lengths from 40—60 cm is useful, particularly if problems are envisaged in finding enough sites to sink them deeply into the stream bottom.
- Paper bags or aluminium pans for drying leaf material
- Materials to process invertebrate samples (e.g. white trays, forceps, etc.)

3. EXPERIMENTAL PROCEDURES

3.1. Experimental Design

The actual design, including frequency, size and relative placement of traps, will be determined by the aim of the project. However, there are several points common to all stream manipulation designs:
1. Identify appropriate reference and experimental stretches and sample all of them *before* manipulation.
2. Ensure that the density of traps is enough to achieve the aim. If the aim is to increase litter mass in the entire channel, then a high proportion of the stream bed needs to be covered by litter traps (for the trap type described above, at least 1 m^{-2} is required). If the aim is simply to increase litter mass in discrete patches around the traps themselves, then a lower density can be used.
3. Sample at appropriate spatial intervals. If stream reach outputs are being measured, then sample immediately downstream of each reach. For within-reach impacts, two alternatives are available. If the aim is to determine the influence of the manipulation on the entire channel, then sample points should be chosen at random in both reference and experimental reaches, with no attempt to include or avoid litter traps in the latter reach. If the aim is to determine the localized influence of the litter traps, then the following is recommended for sampling: random points in the reference stretch; randomly chosen litter traps; random points in the experimental stretch between traps.
4. If multiple sampling dates are used, then sampling the same trap on consecutive dates is best avoided unless dates are at least several months apart.

3.2. Litter Trap Deployment

1. Hammer steel poles into the river bed in pairs, 15—18 cm apart and orientated perpendicular to the flow relative to each other.
2. Carefully thread the mesh rectangle onto the poles and push down until its base is flush with the river bed (Fig. 3.1).
3. Poles should protrude as little as possible above the mesh and the entire trap is most efficient if it breaks the surface slightly at normal flow, thereby capturing detritus floating at all depths.

3.3. Sampling and Sample Processing

1. Trapped litter and associated invertebrates can be sampled with a standard Surber-type sample. If sampling litter traps individually, the Surber sampler is carefully placed over the trap and its associated leaf pack. Then the leaf pack is removed and placed into a bag before sampling the exposed bed in the normal way. The plastic mesh of the trap can be removed and washed during this procedure, then replaced. It is normally instructive to retain the leaf pack and the bed sample as separate samples.
2. Benthic organic matter accumulated between litter traps may also be sampled with a Surber sampler, as may benthic invertebrates.
3. Litter and other benthic organic matter can be processed, dried and ashed, and its weight determined as described in Chapter 1.

Figure 3.1. Newly placed litter traps, at a density of 1 per m². Photo M. Dobson.

4. FINAL REMARKS

The small-scale manipulations described above have been run effectively for several years, with sampling intervals separated by months (Dobson et al. 1995). However, aggregation of leaf litter and animals can occur over a few days or even hours, so more frequent sampling is possible.

The trap described above is small enough to be completely enclosed by a Surber sampler, so traps can be sampled individually to examine small-scale patterns of retention. If, however, the aim is to determine output from an entire reach – for example, aquatic hyphomycete spore concentration or nutrient concentrations in stream water – then using fewer but larger traps will be more practical. The procedures above can be scaled up using, for example, logs that span a large proportion or the entire stream channel (e.g. Smock et al. 1989, Pretty & Dobson 2004; Fig. 3.2).

Figure 3.2. Use of wooden logs as litter traps, as used by Pretty & Dobson (2000). Note that each piece of wood is held in place by four steel poles, arranged in pairs upstream and downstream and each bent over the log at the top to stop it from being entrained by high water flows. Photo M. Dobson.

If the discharge regime of the river fluctuates greatly, and particularly if bed movement occurs during high flow events, the litter traps will act as sediment traps and will eventually fill in. This phenomenon needs to be closely monitored during long term studies. In extreme cases, the traps will initiate development of a series of small islands.

Reducing rather than enhancing retention would initially involve an identification of the major retention structures in the channel, and then their systematic removal (e.g. Wallace et al. 1999, Díez et al. 2000). If retention is mainly by woody debris, then this is straightforward clearance of wood from the channel. If it is cobbles or river bank features such as trailing roots, then removal can be difficult. However, natural retention is generally low in such streams.

Depending upon the source of leaf litter, reduction is also possible by stop-netting upstream of the experimental area, although such nets need constant vigilance as a large mass of debris upstream can quickly build up and may cause them to break. For small-scale projects, small stop-nets can be placed to create localized patches of reduced retention (Fig. 3.3). A range of reach-level measurements (e.g. litter decomposition in mesh bags as described in Chapter 6) can be combined with manipulations of stream retentiveness.

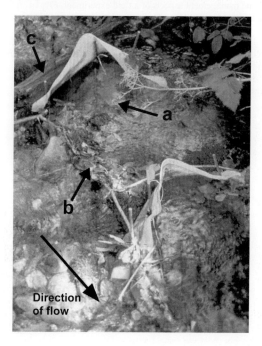

Figure.3.3. Small exclusion nets, designed to reduce inputs of leaf litter from upstream transport to localized patches, in order to monitor the influence of benthic detritus upon leaf decay rates in mesh bags. Note the mesh bag at (a), the unwanted build up of leaf litter at (b), and the large piece of wood caught by the trap at (c). Photo M. Dobson.

5. REFERENCES

Díez, J.R., Larrañaga, S., Elosegi, A. & Pozo, J. (2000). Effect of removal of wood on streambed stability and retention of organic matter. *Journal of the North American Benthological Society*, 19, 621-632.

Dobson M., Hildrew, A.G., Orton, S. & Ormerod, S.J. (1995). Increasing litter retention in moorland streams: ecological and management aspects of a field experiment. *Freshwater Biology*, 33, 325-337.

Pretty, J.L. & Dobson, M. (2004). The response of macroinvertebrates to artificially enhanced detritus levels in plantation streams. Hydrology and Earth System Sciences, 8, 550-559.

Smock, L.A., Metzler, G.W. & Gladden, J.E. (1989). Role of debris dams in the structure and functioning of low gradient headwater systems. *Ecology*, 70, 764-775.

Wallace, J.B., Eggert, S.L., Meyer, J.L. & Webster, J.R. (1999). Effects of resource limitation on a detrital-based ecosystem. *Ecological Monographs*, 69, 409-442.

CHAPTER 4

COARSE BENTHIC ORGANIC MATTER

JESÚS POZO & ARTURO ELOSEGI

*Departamento de Biología Vegetal y Ecología, Facultad de Ciencia y Tecnología
Universidad del País Vasco, C.P. 644, 48080 Bilbao, Spain.*

1. INTRODUCTION

Coarse particulate organic matter (CPOM) is the primary energetic basis of communities in forest streams (Hall et al. 2000). Riparian forests in particular provide streams with substantial amounts of this material (e.g. Cummins et al. 1983, Webster et al. 1995, Abelho 2001), although primary production within the stream can be an additional source of energy. The major components of CPOM are wood, intact and fragmented leaves, and fruits and flowers, with leaves generally dominating in terms of both absolute amount and regularity of input (Fisher & Likens 1972, Pozo et al. 1997, Abelho 2001).

Once in the stream, the retention of CPOM depends, among other factors, on the hydrologic regime, channel morphology, and presence of debris dams. Together with the vegetation canopy, these factors can control the accumulation of coarse benthic organic matter (CBOM), generally resulting in a highly patchy distribution (Smock 1990).

Temporal variability of CBOM can also be high. It is mainly due to leaf-fall phenology and to the hydrologic regime (e.g. Molinero & Pozo 2002). In temperate deciduous forest streams, leaf fall peaks in autumn and early winter and this can be reflected in the amounts of CBOM in the stream at this time (Iversen et al. 1982, Bärlocher 1983). Often, however, the input of autumn-shed leaves correlates poorly with the dynamics of CBOM under changing discharge conditions. Only when high inputs coincide with a period of low flow, can an increase in CBOM in the stream be expected (Pozo et al. 1997). Lowest values are often found during spring and summer. Clearly, studies on CBOM in streams should consider variability across both spatial and temporal scales.

Disturbances of the riparian vegetation such as clear-cutting and plantations of exotic species can alter the quantity, quality and temporal and spatial distribution of inputs and storage in streams (Webster et al. 1990, Graça et al. 2002). However, the high variability of values reported in the literature (Abelho 2001) is also partly due

25

*M.A.S. Graça, F. Bärlocher & M.O. Gessner (eds.), Methods to Study Litter Decomposition:
A Practical Guide,* 25 – 32.
© 2005 *Springer. Printed in The Netherlands.*

to the use of different sampling methods and size fractionation in different studies. Both should be taken into account when comparing data, such as those shown in Table 4.1.

Table 4.1. Coarse benthic organic matter (excluding large wood) from selected streams.

Location	Latitude	Vegetation	Stream order	CBOM (g AFDM m^{-2})	Reference
Alaska, USA	65 °N	Taiga	1	3	1
			2	8	1
Denmark	56 °N	Beech forest	1	135*	2
Quebec, Canada	50 °N	Spruce and deciduous forest	1	968	3
			2	317	3
			5	456	3
Switzerland	47 °N	Deciduous forest	3	27**	4
Oregon, USA	45 °N	Coniferous forest	1	1012—5117	5
			3	388	5
			5	61	5
New Hampshire, USA	44 °N	Deciduous forest	2	509***	6
Spain	43 °N	Deciduous forest	1	60	7
		Deciduous forest	3	20	7
		Eucalyptus and deciduous forest	3	12	7
		Eucalyptus plantation	1	62—75	8
Pennsylvania, USA	40 °N	Agricultural and woodland	3	118	9
Virginia, USA	37 °N	Mixed forest	1	1730	10
North Carolina, USA	35 °N	Deciduous forest	1	391	11
		Logged deciduous forest	2	286	11
Arizona, USA	33 °N	Desert scrub	5	5	12
South Africa	33 °S	Fynbos biome	2	19—32**	13
Victoria, Australia	37 °S	Eucalyptus forest	4	105	14

*Leaves only; **Converted from dry mass (DM), assuming that AFDM = 0.9 × DM; ***Converted from kcal, assuming 10 kcal = 1 g C = 2 g AFDM); 1= Irons & Oswood (1997); 2 = Iversen et al. (1982); 3 = Naiman & Link (1997); 4 = Bärlocher (1983); 5 = Webster & Meyer (1997); 6 = Fisher & Likens (1972); 7 = González & Pozo (1996); 8 = Molinero & Pozo (2002); 9 = Newbold et al. (1997); 10 = Smock (1990); 11 = Webster et al. 1990; 12 = Jones et al. (1997); 13 = King et al. (1987); 14 = Treadwell et al. (1997).

CBOM estimates have been based on a variety of methods: random sampling of the wetted channel (González & Pozo 1996), sampling of transects either at random (Golladay et al. 1989) or at regular intervals along the stream (Wallace et al. 1995), and stratified random sampling (Mullholand 1997). Samples are usually taken with a Surber-type sampler or a power-vacuum assisted, cylindrical corer. This chapter describes a method to estimate the amounts of CBOM stored in small streams with a

Surber-type sampler. Potential applications of this method include assessments of differences in CBOM among similar-sized streams experiencing different degrees of disturbance of the riparian vegetation. In addition, relationships of CBOM storage with the retention capacity of streams (see chapter 2), their flow regime, phenology of allochthonous inputs or other temporal changes, may be explored.

2. SITE SELECTION AND EQUIPMENT

2.1. Site Selection

CBOM storage is most readily estimated in small forested headwater streams (stream order 1 or 2). Choose an accessible stream segment as homogeneous as possible in terms of riparian vegetation, geomorphology and substrate. A reach of 100 m will normally be sufficient.

2.2. Equipment and Material

- Aluminium trays
- Balance
- Bucket
- Crucibles
- Desiccator
- Drying oven
- Freezer
- Labelled plastic bags
- Modified Surber sampler (sampling surface of 0.25 m^2; mesh size of 1 mm)
- Muffle furnace
- Set of nested sieves (1-mm and 1-cm mesh sizes)
- Plastic trays
- Random number table
- Small shovel
- Tape measure
- Tongs

3. EXPERIMENTAL PROCEDURES

3.1. Sampling

1. Use a random number table and tape measure to choose 5 points along the selected stream reach at random. Establish a 0.5-m wide transect across the stream from bank to bank (including dry parts of the channel).
2. Note the width of the channel in each transect.
3. If the transect includes a dry section the following procedures must be used:

(a) On the dry section, collect all the substrate with a small shovel to a depth of 5 cm when possible. Eliminate large inorganic substrates before putting the collected material in a set of nested sieves (sieve of 1-cm mesh size on top of a sieve of 1-mm mesh size). Rinse the sample with stream water and eliminate, as much as possible, all inorganic materials and wood pieces >1 cm in diameter that are retained by the 1-cm sieve. Transfer the rest to a labelled plastic bag. Transfer the material retained by the 1-mm sieve to the same plastic bag.

(b) In the submerged section, collect the CBOM with the modified Surber sampler. Be sure to disturb the substrate to a depth of at least 5 cm in a standardized fashion. Transfer the material retained by the net of the Surber sampler to the nested sieves and discard the inorganic substrates and wood pieces >1 cm in diameter before putting the rest of the sample in a labelled plastic bag.

4. Proceed in the same way with the other transects of the stream segment.
5. Carry the collected CBOM to the laboratory. If samples cannot be processed within the next few days, freeze them as soon as possible.
6. In temporally extended studies, samples can be collected twice a month during periods of heavy litter fall, and monthly thereafter. Avoid sampling the same transects repeatedly. Direct flow measures or discharge records from the nearest gauging station may be useful when interpreting temporal changes in CBOM.

3.2. Laboratory Procedures

1. Remove attached sand and silt from the CBOM collected in the nested sieves by rinsing with tap water. Transfer all the organic matter to a plastic tray.
2. Sort CBOM into the following categories: leaves, twigs, bark, fruits and flowers, and debris (unidentifiable fragments >1 mm). Remove any remaining branches >1 cm in diameter. Sort leaves by species. Put each CBOM category in a separate pre-weighed aluminium tray.
3. Dry at 40—50 °C.
4. Let the samples cool in a desiccator and weigh.
5. Transfer the material of each tray to a pre-weighed crucible (a weighed subsample can be used if the amount of CBOM is high).
6. Put the crucibles in the muffle furnace and ash at 500 °C for 4 h.
7. Let the crucibles cool in a desiccator and weigh.

3.3. Calculations

1. To calculate the area (m^2) of each transect, multiply the respective channel width (m) by 0.5 m (width of sampled transect).
2. Calculate the ash-free dry mass (AFDM) of each CBOM category by subtracting the ash weight from dry weight.

3. Add the values of all categories in a sample to obtain the total AFDM.
4. Divide the results by the area of the respective transect to express them in terms of g AFDM m^{-2}.

4. FINAL REMARKS

Mosses, when present, can easily be removed from the remaining CBOM, allowing estimates of their respective contributions. Since a considerable fraction of the total CBOM can be found below the streambed surface (Cummins et al. 1983), it is critical for most applications to take into account this buried material.

   If the main concern is invertebrate food availability, sampling may be restricted to the wetted channel. If the objective is to construct an organic matter budget (see Chapter 7), the whole channel transect has to be sampled.

5. REFERENCES

Abelho, M. (2001). From litterfall to breakdown in streams: a review. *The ScientificWorld*, 1, 656-680.

Bärlocher, F. (1983). Seasonal variation of standing crop and digestibility of CPOM in a Swiss Jura stream. *Ecology*, 64, 1266-1272.

Cummins, K.W., Sedell, J.R., Swanson, F.J., Minshall, F.W., Fisher, S.G., Cushing, C.E., Peterson, R.C. & Vannote, R.L. (1983). Organic matter budgets for stream ecosystems: problems in their evaluation. In: J.R. Barnes & G.W. Minshall (eds.), *Stream Ecology. Application and testing of general ecological theory* (pp. 299-353). Plenum Press. New York

Fisher, S.G. & Likens, G.E. (1972). Stream ecosystem: organic energy budget. *BioScience*, 22, 33-35.

Golladay, S.W., Webster, J.R. & Benfield, E.F. (1989). Changes in stream benthic organic matter following watershed disturbance. *Holarctic Ecology*, 12, 96-105.

González, E. & Pozo, J. (1996). Longitudinal and temporal patterns of benthic coarse particulate organic matter in the Agüera stream (northern Spain). *Aquatic Sciences*, 58, 355-366.

Graça, M.A.S., Pozo, J., Canhoto, C. & Elosegi, A. (2002). Effects of *Eucalyptus* plantations on detritus, decomposers, and detritivores in streams. *The ScientificWorld*, 2, 1173-1185.

Hall, R.O. Jr., Wallace, J.B. & Eggert, S.L. (2000). Organic matter flow in stream food webs with reduced detrital resource base. *Ecology*, 81, 3445-3463.

Irons, J.G. & Oswood, M.W. (1997). Organic matter dynamics in 3 subarctic streams of interior Alaska, USA. In: J.R. Webster & J.L. Meyer (eds.), Stream organic matter budgets. *Journal of the North American Benthological Society*, 16, 3-161, pp. 23-28.

Iversen, T., Thorup, J. & Skriver, J. (1982). Inputs and transformation of allochthonous particulate organic matter in a headwater stream. *Holarctic Ecology*, 5, 10-19.

Jones, J.B., Schade, J.D., Fisher, S.G. & Grimm, N.B. (1997). Organic matter dynamics in Sycamore Creek, a desert stream in Arizona, USA. In: J.R. Webster & J.L. Meyer (eds.), Stream organic matter budgets. *Journal of the North American Benthological Society*, 16, 3-161, pp. 78-82.

King, J.M., Day, J.A., Davies, B.R. & Hensall-Howard, M.P. (1987). Particulate organic matter in a mountain stream in the South-western Cape, South Africa. *Hydrobiologia*, 154: 165-187.

Molinero, J. & Pozo, J. (2002). Impact of eucalypt plantations on the benthic storage of coarse particulate organic matter, nitrogen and phosphorus in small streams. *Verhandlungen der Internationalen Vereinigung für Limnologie*, 28, 540-544.

Mulholland, P.J. (1997). Organic matter dynamics in the West Fork of Walker Branch, Tennessee, USA. In: J.R. Webster & J.L. Meyer (eds.), Stream organic matter budgets. *Journal of the North American Benthological Society*, 16, 3-161, pp. 61-67.

Naiman, R.J. & Link, G.L. (1997). Organic matter dynamics in 5 subarctic streams, Quebec, Canada. In: J.R. Webster & J.L. Meyer (eds.), Stream organic matter budgets. *Journal of the North American Benthological Society*, 16, 3-161, pp. 33-39.

Newbold, J.D., Bott, T.L., Kaplan, L.A., Sweeney, B.W. & Vannote, R.L. (1997). Organic matter dynamics in White Clay Creek, Pennsylvania, USA. In: J.R. Webster & J.L. Meyer (eds.), Stream organic matter budgets. *Journal of the North American Benthological Society*, 16, 3-161, pp. 46-50.

Pozo, J., González, E., Díez, J.R. & Elosegi., A. (1997). Leaf-litter budgets in two contrasting forested streams. *Limnetica*, 13, 77-84.

Smock, L.A. (1990). Spatial and temporal variation in organic matter storage in low-gradient, headwater streams. *Archiv für Hydrobiologie*, 118, 169-184.

Treadwell, S.A., Campbell I.C. & Edwards, R.T. (1997). Organic matter dynamics in Keppel Creek, southeastern Australia. In: J.R. Webster & J.L. Meyer (eds.), Stream organic matter budgets. *Journal of the North American Benthological Society*, 16, 3-161, pp. 58-61.

Wallace, J.B., Whiles, M.R., Eggert, S., Cuffney, T.F., Lugthart, G.J. & Chung. K. (1995). Long-term dynamics of coarse particulate organic matter in three Appalachian Mountain streams. *Journal of the North American Benthological Society*, 14, 217-232.

Webster, J.R. & Meyer, J.L. (1997). Stream organic matter budgets-introduction. In: J.R. Webster & J.L. Meyer (eds.), Stream organic matter budgets. *Journal of the North American Benthological Society*, 16, 3-161, pp. 5-13.

Webster, J.R., Golladay, S.W., Benfield, E.F., D'Angelo, D.J. & Peters, G.T. (1990). Effects of forest disturbance on particulate organic matter budgets of small streams. *Journal of the North American Benthological Society*, 9, 120-140.

Webster, J.R., Wallace, J.B. & Benfield, E.F. (1995). Organic processes in streams of the Eastern United States. In: C.E. Cushing, K.W. Cummins & G.W. Minshall (eds.), *Ecosystems of the World 22: River and Stream Ecosystems* (pp. 117-187). Elsevier. Amsterdam.

CHAPTER 5

LEACHING

FELIX BÄRLOCHER

63B York Street, Department of Biology, Mt. Allison University, Sackville, NB, Canada E4L 1G7.

1. INTRODUCTION

The decomposition of autumn-shed leaves has traditionally been subdivided into three more or less distinct phases: leaching, microbial colonization and invertebrate feeding (Petersen & Cummins 1974, Gessner et al. 1999). Leaching is defined as the abiotic removal of soluble substances, among them phenolics, carbohydrates and amino acids (for analyses of these compounds, see Chapters 10, 11 and 14). It is largely completed within the first 24—48 h after immersion in water, and results in a loss of up to 30% of the original mass, depending on leaf species. Gessner & Schwoerbel (1989) showed that no such rapid leaching loss can be observed when fresh, rather than pre-dried, alder and willow leaves are used. Fungal colonization proceeded more slowly on fresh than on pre-dried alder and willow leaves (Bärlocher 1991, Chergui & Pattee 1992), dynamics of chemical leaf constituents differed between fresh and pre-dried leaves during subsequent decomposition (Gessner 1991), but no effects on invertebrate colonization have been observed (Chergui & Pattee 1993, Gessner & Dobson 1993). In a survey of 27 leaf species, drying significantly changed the magnitude of leaching in a majority of cases (Taylor & Bärlocher 1996), although the direction of change was variable among species with drying actually decreasing leaching in several cases. Some representative data are listed in Table 5.1.

Changes in types and amounts of compounds retained by leaves may affect their breakdown rate by selectively stimulating or inhibiting colonization by aquatic microorganisms (Bengtsson 1983, 1992) and by modifying palatability to leaf-eating invertebrates (review in Bärlocher 1997). In addition, they will influence the dynamics of the dissolved organic matter pool in the water column, its flocculation into solid particles (Bärlocher et al. 1989), and its entrapment and processing at liquid-solid interfaces (Armstrong & Bärlocher 1989a,b, Meyer et al. 1998, Allan 1995).

M.A.S. Graça, F. Bärlocher & M.O. Gessner (eds.), Methods to Study Litter Decomposition: A Practical Guide, 33 – 36.

Table 5.1. Percentages of mass losses of fresh and dried leaves over 48 h in distilled water. Mean±SD. <, loss significantly greater in dried leaves; =, no significant difference; >, loss significantly greater from fresh leaves. Data from Taylor & Bärlocher (1996).

| | Leaf mass loss (% dry mass) | | |
Leaf species	Fresh		Dried
Acer saccharum	15.2±7.9	<	21.4±7.6
A. negundo	14.7±3.3	<	30.5±2.1
A. circinatum	6.3±5.2	<	23.7±1.5
A. rubrum	16.6±9.2	=	24.5±8.2
Fagus grandifolia	5.1±7.0	=	7.4±6.6
A. macrophyllum	10.2±6.7	>	5.9±0.9
Betula papyrifera	15.1±2.4	>	11.7±1.3

In some areas, the yearly leaf fall may overlap with the first night frosts. Freezing living or senescent leaves can have a similar effect as drying them: it may damage cell membranes, which generally accelerates leaching (Bärlocher 1992). In other areas, leaf senescence may coincide with hot, dry weather, and leaves may dry on the tree.

The method described here allows assessing how drying leaves influences leaching. Freshly collected, non-dried leaves (fresh leaves) and leaves that are dried after collection (dried leaves) are exposed in fine-mesh bags (to prevent access by macroinvertebrates) in a stream. After four days, the remaining mass is measured. During this early period of decomposition, leaching generally predominates. If drying significantly increases leaching, we expect higher losses in dried leaves. If desired, identically treated leaves can be examined for colonization by aquatic hyphomycetes. To study the temporal course of leaching losses in greater detail, leaf bags should be prepared to allow daily samples for an extended period of seven days. Or, leaves can be submerged in distilled or stream water in the laboratory, and daily samples can be taken (Gessner & Schwoerbel 1989, Taylor & Bärlocher 1996).

2. EQUIPMENT, CHEMICALS AND SOLUTIONS

2.1. Equipment and Material

- Oven (40—50 °C)
- Leaves of *Alnus glutinosa* or other species
- Litter bags (10 x 10 cm, mesh size 0.5 or 1 mm)
- Plastic labels
- Balance (±1 mg)

3. EXPERIMENTAL PROCEDURES

3.1. Sample Preparation

1. Dry leaves: collect leaves from a single tree by gently shaking branches and collecting fallen leaves. Dry for 2 days at 40—50 °C to constant mass. Randomly select 2— 3 leaves and weigh to the nearest mg. Moisten leaves to avoid breakage by placing the leaves in a small tray and spraying them with water (avoid highly chlorinated tap water), and place them in a litter bag (see Chapter 6). Label the bag. Prepare a total of 20 bags.
2. Fresh leaves: harvest leaves from the same tree. Return them to laboratory in a cool, closed container. Randomly select 2—3 leaves, weigh them, and place them in a litter bag. Label the bag. Prepare a total of 20 bags. To determine wet mass/dry mass ratio of fresh leaves, individually weigh 20 fresh leaves, dry them, and weigh them again.

3.2. Experiment

1. Expose all bags in a stream.
2. Recover all bags after 4 days.
3. Rinse leaves under running tap water, dry at 40—50 °C to constant mass, and weigh them.
4. Express mass loss as percentage of original leaf mass.

3.3. Statistical Analysis

Mass losses of fresh and dried leaves can be compared with a *t*-test or a permutation test (see Chapter 43). Since some values are likely to be below 20%, normal distribution cannot be assumed, and arcsine transformation of proportion p is advisable before applying a standard *t*-test ($p' = \arcsin \sqrt{p}$).

For the permutation test, we assume that the values for fresh and dried leaves belong to the same population (H_0, null hypothesis). We therefore pool all values. Next, we randomly divide the 40 values into two groups of 20. We determine the difference between mean mass losses of the two groups. We do this 10000 times and plot the distribution of the differences. Next, we determine the actual difference between the original data from fresh and dried leaves. How "extreme" is it? If it is at least as extreme as 5% of the population of differences based on the permutated data (this corresponds to $p \leq 0.05$), we reject the null hypothesis.

4. REFERENCES

Allan, J.D. (1995). *Stream Ecology. Structure and Function of Running Waters*. Chapman & Hall. London.

Armstrong, S.M. & Bärlocher, F. (1989a). Adsorption and release of amino acids from epilithic biofilms in streams. *Freshwater Biology*, 22, 153-159.

Armstrong, S.M. & Bärlocher, F. (1989b). Adsorption of three amino acids to biofilms on glass-beads. *Archiv für Hydrobiologie*, 115, 391-399.

Bärlocher, F. (1991). Fungal colonization of fresh and dried leaves in the River Teign (Devon, England). *Nova Hedwigia*, 52, 349-357.

Bärlocher, F. (1992). Effects of drying and freezing autumn leaves on leaching and colonization by aquatic hyphomycetes. *Freshwater Biology*, 28, 1-7.

Bärlocher, F. (1997). Litter breakdown in rivers and streams. *Limnética*, 13, 1-11.

Bärlocher, F., Tibbo, P.G. & Christie, S.H. (1989). Formation of phenol-protein complexes and their use by two stream invertebrates. *Hydrobiologia*, 173, 243-249.

Bengtsson, G. (1983). Habitat selection in two species of aquatic hyphomycetes. *Microbial Ecology*, 9, 15-26.

Bengtsson, G. (1992). Interactions between fungi, bacteria and beech leaves in a stream microcosm. *Oecologia*, 89, 542-549.

Chergui, H. & Pattee, E. (1992). Processing of fresh and dry *Salix* leaves in a Moroccan river system. *Acta Oecologica*, 13, 291-298.

Chergui, H. & Pattee, E. (1993). Fungal and invertebrate colonization of *Salix* fresh and dry leaves in a Moroccan river system. *Archiv für Hydrobiologie*, 127, 57-72.

Gessner, M.O. (1991). Differences in processing dynamics of fresh and dried leaf litter in a stream ecosystem. *Freshwater Biology*, 26, 387-398.

Gessner, M.O. & Schwoerbel, J. (1989). Leaching kinetics of fresh leaf-litter with implications for the current concept of leaf-processing in streams. *Archiv für Hydrobiologie*, 115, 81-90.

Gessner, M.O. & Dobson, M., (1991). Colonization of fresh and dried leaf-litter by lotic macroinvertebrates. *Archiv für Hydrobiologie*, 127, 141-149.

Gessner, M.O., Chauvet, E. & Dobson, M. (1999). A perspective on leaf litter breakdown in streams. *Oikos*, 85, 377-384.

Meyer, J.L., Wallace, J.B. & Eggert, S.L. (1998). Leaf litter as a source of dissolved organic carbon in streams. *Ecosystems*, 1, 240-249.

Petersen, R.C. & Cummins, K.W. (1974). Leaf processing in a woodland stream. *Freshwater Biology*, 4, 343-368.

Taylor, B.R. & Bärlocher, F. (1996). Variable effects of air-drying on leaching losses from tree leaf litter. *Hydrobiologia*, 325, 173-182.

CHAPTER 6

LEAF MASS LOSS ESTIMATED BY LITTER BAG TECHNIQUE

FELIX BÄRLOCHER

63B York Street, Department of Biology, Mt. Allison University, Sackville, NB, Canada E4L 1G7.

1. INTRODUCTION

Leaf litter is a dominant component of coarse particulate organic matter in streams, and its decomposition has received considerable attention (Webster & Benfield 1986, Allan 1995, Gessner et al. 1999). Gessner & Chauvet (2002) proposed using leaf litter breakdown to evaluate functional stream integrity. In order to increase the sensitivity and robustness of the assay, 'noise' due to non-standardized procedures has to be minimized. Many studies have used pre-dried leaves or leaf disks enclosed in litter bags. Several aspects of this approach have been criticized as introducing artificial modifications of the natural process (Petersen & Cummins 1974, Wieder & Lang 1982, Boulton & Boon 1991, Bärlocher 1997). Mass loss in litter bags (1 mm mesh size) resembles that of loose, naturally entrained leaves in depositional zones, while mass loss in litter packs (leaves tied together and tethered in streams) is close to that of loose leaves in riffle areas (Cummins et al. 1980). On the other hand, the use of litter bags with different mesh sizes allows size-selective exclusion of macro-consumers.

The mass loss of leaves and needles as a function of time is most often approximated by an exponential decay model:

$$M_t = M_0 \cdot e^{-kt} \qquad (6.1)$$

where M_t = mass at time t; M_0 = mass at time 0; k = exponential decay coefficient; and t = time in days. Based on their daily decay coefficient, leaves have been classified as "fast" ($k > 0.01$), "medium" ($k = 0.005—0.001$) and "slow" ($k < 0.005$) based on extensive work in a stream in Michigan (Petersen & Cummins 1974). However, the decomposition rate of a given leaf species can vary greatly among streams (Suberkropp & Chauvet 1995), suggesting that Petersen & Cummins' (1974)

M.A.S. Graça, F. Bärlocher & M. Gessner (eds.), Methods to Study Litter Decomposition: A Practical Guide, 37 – 42

classification has limitations when applied uncritically. A few typical values of k, with number of days required for 50% mass loss, are listed in Table 6.1.

Table 6.1. Daily decay rates, k, and number of days to reach 50% mass loss $(=T_{50})$, of selected leaf species (data from Petersen & Cummins 1974, based on leaf packs).

Category	Species	k	T_{50}
Fast	Fraxinus americana	0.0120	58
	Tilia americana	0.0175	40
Medium	Carya glabra	0.0089	78
	Salix lucida	0.0078	89
Slow	Fagus grandifolia	0.0025	277
	Quercus alba	0.0022	315

The exponential decay equation is typically converted to a linear form before regression is performed.

$$\ln[M_t] = \ln[M_0 \cdot e^{-kt}] = \ln[M_o] - kt \tag{6.2}$$

rewritten as

$$Y = a + bX \tag{6.3}$$

Y is the dependent variable, corresponding to M_t. The independent variable X equals time in days. The linear regression procedure, which minimizes the sum of squares, determines the slope b (equals decay coefficient k) and the intercept a (calculated mass at time 0, which should be close to 100%). Most computer programs will also calculate R^2. This indicates how much of the variance among the data is due to the linear relationship between the X and Y values. For example, a value of 0.95 corresponds to 95%.

Linear regression calculations are only valid when the experimental uncertainty of replicate Y values is not related to the values of X or Y (Zar 1998). This is not usually the case after data transformation, which tends to enhance errors associated with small Y values. These points will be emphasized by linear regressions and points with large Y values will be relatively ignored. Thus, linearizing transformations is not an ideal procedure because it distorts experimental errors (Motulsky & Ransnas 1987, Motulsky & Christopoulos 2003).

The alternative is nonlinear regression analysis. This is defined as fitting data to any selected equation. As with linear regression, nonlinear regression procedures determine values of the parameters that minimize the sum of the squares of the distances of the data points to the curve. The approach is only appropriate when the experimental uncertainty is normally distributed, and not related to the values of X or Y. Nonlinear regression (or curve-fitting) must be solved iteratively, rather than analytically, and an initial estimate of each parameter must usually be provided.

During the fitting procedure, these values are modified to increasingly improve the fit (lower the sum of squared deviations) of the curve to the data. These iterations are continued until additional improvements are negligible.

Often two sets of data are fitted to the same model, and the question is whether the two sets of data differ significantly. For example, do eucalypt and alder leaves decay at significantly different rates? A good introduction to this topic can be found at http://www.graphpad.com/curvefit. The recommended approach is to repeat the experiment several times and compare the resulting estimates of the parameter k with a t-test, which compares a difference with the standard error of that difference. This method is labour-intensive and statistically conservative; the calculated p value may be too high. If the experiment has only been done once, the best-fit value of two groups can still be compared with a t-test by using the standard error reported by the curve-fitting program. This again assumes normal error distribution, which is approximately true for the exponential decay equation.

A commonly used approach is to analyze the two data sets separately as well as simultaneously. The question then is whether the separate fits are significantly better than the pooled fit. This method, known as Analysis of Covariance or ANCOVA, is again strictly valid only for a linear relationship between X and Y. The details of this test can be found in Zar (1998). Another approach is based on Akaike's Information Criterion (AIC), which answers the following questions: Which model is more likely to have generated the data? How much more likely? The theory behind AIC is quite difficult; it combines maximum likelihood, information theory, and the concept of information entropy. Fortunately, the computations and interpretation of the results are straightforward (Motulsky & Christopoulos 2003).

Finally, two (or more) data sets can be compared by permutation or randomization tests. The first step is to define a test statistic S, e.g., the difference between several estimates of the k values of alder and eucalypt. Calculate the value of S for the original data set. All values are then pooled and randomly assigned to alder or eucalypt. The difference is calculated. This is repeated many times, giving the distribution of all possible values of S. The final question is: How extreme is the S of the original data compared to all possible values? If it is more extreme than 5% (or 1%) of the population, H_0 is rejected. If we do not want to make the assumption of linear decay, we can choose another test statistic. For example, we can simply add up all data (% remaining mass on all sampling dates) separately for the two leaf species, and determine the difference. The null hypothesis is that low and high values are randomly distributed between alder and eucalypt. We again pool the data, randomly distribute them between alder and eucalypt, and calculate the new value of S. This is repeated many times, and we determine how extreme the original S is.

Sometimes, the single exponential model is clearly inappropriate (Wieder & Lang 1982). When the leaf consists of two clearly defined components decaying at different rates, a double exponential equation gives a better fit. When decomposition does not appear to proceed beyond a certain point, an asymptotic model provides a better fit (Sridhar et al. 2002). How does one decide which model gives a better fit? (Zen Koan of Statistics: the person with one watch knows what time it is; the person with two watches is never sure.) A simple comparison of sum of squares (or R^2) values is inappropriate, since a curve with more parameters nearly always has a lower sum of squares because it has more inflection points (Kvålseth 1985). The question is whether this decrease is worth the "cost" of the additional variables, whose inclusion in the model results in a loss of degrees of freedom. Two approaches are commonly used: an F test (extra sum-of-squares test), or Akaike's Information Criterion. The appropriate procedures for making the decision are described in Motulsky & Christopoulos (2003).

2. EQUIPMENT, CHEMICALS AND SOLUTIONS

2.1. Equipment and Material

- Autumn-shed leaves (fresh and/or air-dried)
- Litter bags (10 x 10 cm, 1-mm mesh size; alternatively, 0.5-mm and 10-mm mesh size)
- Plastic labels (DYMO or laser printed numbers on transparencies)
- Drying oven (40—50°C)
- Balance (±1 mg precision)
- Statistics program or calculator

3. EXPERIMENTAL PROCEDURES

3.1. Sample Preparation

1. Leaves are collected and prepared as in Chapter 5 (Leaching). Depending on objectives, mass loss rates of different leaf species, of leaves in different mesh size bags or leaves placed in different rivers can be compared.
2. Place 2—3 preweighed leaves (or 3—4 g) in each labelled litter bag.
3. Prepare a sufficient number of bags to allow 4—6 replicates per sampling date plus two extra sets, one set to convert air-dry mass to oven-dry mass and another set to determine handling losses (see below).

3.2. Exposure of Litter Bags

1. Anchor leaf bags to the stream bed with bricks, steel pegs, etc. Depending on objectives, all bags may be placed in riffle or pool areas. Care must be taken not to place too many bags close to each other, because this may drastically change current patterns and thereby affect colonization by microorganisms and invertebrates. Alternatively, leaf bags may be attached to rebars anchored to the stream bed.

3.3. Recovery of Bags and Analysis

2. To correct air-dry mass of all leaf bags for humidity, dry a first set of bags at 40—50 °C for 2 days (or until mass remains constant) and weigh. Calculate an average correction factor, D.
3. D = (oven-dry mass) / (air-dry mass).
4. The initial oven-dry mass of each leaf pack brought to the streams is estimated by multiplying the measured air-dry mass by the average correction factor, D.
5. A second set of bags should be recovered immediately upon exposure in the stream. This allows an estimate of losses due to initial handling.
6. Subsequent samples are taken according to a preplanned schedule, for example after 3, 7, 14, 21, 28, and 35 days. With slowly decomposing leaves, an extended sampling schedule (up to 3 months or even longer) may be necessary.
7. Rinse leaves under running tap water, dry to constant mass and weigh.

3.4. Statistical Analysis

1. Express mass loss as percentage of original mass after correction for (1) humidity and (2) handling (100% = mass after corrections).
2. Run a regression analysis. (a) A linear regression with the original data (Time = independent variable; % Mass remaining = dependent variable). (b) A non-linear curve-fitting program (exponential decay; in most programs, you will have to type in the equation, and provide initial estimates of the parameters that are to be determined (in the exponential decay model, provide an estimate of a and k). (c) Transform [% Mass remaining] to ln[% Mass remaining] and run a linear regression (Time = independent variable; ln[% Mass remaining] = dependent variable).
3. Generally, the differences in slope estimates among the three models are small, but the non-linear curve-fit often provides the best estimate for the intercept (which, by definition, is 100%).
4. If data from more than one series have been collected (e.g., 2 or more leaf species), an analysis of covariance using Time as covariate can be run. This can be done in some computer programs, or as described in Zar (1998). Provided the estimated initial leaf mass of the two series are similar, the decay coefficients are significantly different if the p value for the interaction between time and series is < 0.05. Alternatively, an appropriate test statistic may be formulated and a corresponding permutation test performed (e.g. with the software Resampling Stats; see also Chapter 43).
5. If a pronounced steep decline during the early phase of decomposition is observed (which may be due to leaching), try fitting the data to a more complex model (e.g., double exponential decay). Motulsky & Christopoulos (2003) can be a useful to decide which model is more appropriate.

4. REFERENCES

Allan, J.D. (1995). *Stream Ecology. Structure and Function of Running Waters*. Chapman & Hall. London.

Bärlocher, F. (1997). Pitfalls of traditional techniques when studying decomposition of vascular plant remains in aquatic habitats. *Limnética*, 13, 1-11.

Boulton, A.J. & Boon, P.I. (1991). A review of the methodology used to measure leaf litter decomposition in lotic environments: time to turn over an old leaf? *Australian Journal of Marine and Freshwater Research*, 42, 1-43.

Cummins, K.W., Spengler, G.L., Ward, G.M., Speaker, R.M., Ovink, R.W., Mahan, D.C. & Mattingly, R.L. (1980). Processing of confined and naturally entrained leaf litter in a woodland stream ecosystem. *Limnology and Oceanography*, 25, 952-957.

Gessner, M.O., Chauvet, E., & Dobson, M. (1999). A perspective on leaf litter breakdown in streams. *Oikos*, 85, 377-384.

Gessner, M.O. & Chauvet, E. (2002). A case for using litter breakdown to assess functional stream integrity. *Ecological Applications*, 12, 498-510.

Kvålseth, T.O. (1985). Cautionary note about R^2. *The American Statistician*, 39, 279-285.

Motulsky, H.J. (1995). *Intuitive Biostatistics*. Oxford University Press. Oxford.

Motulsky, H.J. & Rasnsnas, L.A. (1987). Fitting curves to data using nonlinear regression: a practical and non-mathematical review. *Journal of The Federation of American Societies for Experimental Biology*, 1, 365-374.

Motulsky, H.J. & Christopoulos, A. (2003). Fitting models to biological data using linear and nonlinear regression. A practical guide to curve fitting. San Diego: GraphPad Software, Inc., www.graphpad.com

Petersen, R.C. & Cummins, K.W. (1974). Leaf processing in a woodland stream. *Freshwater Biology*, 4, 343-368.

Sridhar, K.R., Krauss, G., Bärlocher, F., Raviraja, N.S., Wennrich, R. & Baumbach, R. (2002). Decomposition of alder leaves in two heavy metal polluted streams in Central Germany. *Aquatic Microbial Ecology*, 26, 73-80.

Wieder, R.K. & Lang, G.E. (1982). A critique of the analytical methods used in examining decomposition data obtained from litter bags. *Ecology*, 63, 1636-1642.

Zar, J.H. (1998). *Biostatistical Analysis*. Prentice-Hall. New York.

CHAPTER 7

COARSE PARTICULATE ORGANIC MATTER BUDGETS

JESÚS POZO

Departamento de Biología Vegetal y Ecología, Facultad de Ciencia y Tecnología
Universidad del País Vasco, C.P. 644, 48080 Bilbao, Spain.

1. INTRODUCTION

During the last three decades great efforts have been dedicated to the study of terrestrial-aquatic linkages, in particular to coarse particulate organic matter (CPOM) of riparian origin and its fate in streams (see Webster et al. 1995 and Abelho 2001 for reviews). A considerable amount of litter entering streams is retained within the channel. Thus local benthic CPOM is related to the riparian vegetation, underlining the strong influence of the terrestrial environment on the energy basis of low-order forest streams (Wallace et al. 1999). In addition, the discharge regime directly affects the stream retention capacity for CPOM (Larrañaga et al. 2003) and thus the availability of organic matter to stream consumers and decomposers. When peaks of litter input coincide with high discharge, downstream displacement of CPOM is favoured, whereas CPOM tends to accumulate in the stream bed when litter inputs coincide with low flow.

A budget is a set of calculations accounting for inputs (I) and outputs (O) in a given ecosystem or part thereof. The general approach to constructing a CPOM budget consists of determining the various inputs (see Chapter 1) and ascertaining outputs (see Chapter 6) and the storage (=standing stock, S; see Chapter 4) of this material (Fig. 7.1). Inputs to a stream (vertical, lateral and from upstream), outputs (downstream transport or export and biological breakdown), and changes in storage should all be measured independently. However, this approach is very time-consuming, and often one or more variables are obtained by addition or subtraction of measured components, taken from the literature, or ignored (Cummins et al. 1983, Minshall 1996).

The general mass balance for CPOM in a stream segment is given by:

$$O = I + \Delta S \qquad (7.1)$$

43

M.A.S. Graça, F. Bärlocher & M.O. Gessner (eds.), Methods to Study Litter Decomposition:
A Practical Guide, 43 – 50
© 2005 *Springer. Printed in The Netherlands.*

where total outputs in a given time period (O) equal total inputs (I) plus the change in standing stock (ΔS) in that time period. In a more detailed form, the standing stock of CPOM at time t (S_t) is given by the standing stock at time 0 (S_0) plus total inputs from t_0 to t (I), minus the losses resulting from downstream transport (O_T) and biological and physical breakdown (O_B):

$$S_t = S_0 + I - O_T - O_B \qquad (7.2)$$

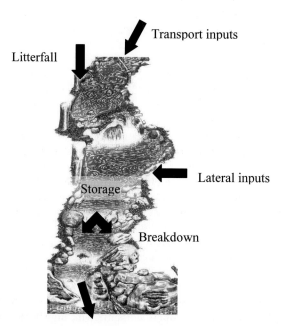

Transport inputs

Litterfall

Lateral inputs

Storage

Breakdown

Transport outputs

Figure 7.1. Components of a coarse particulate organic matter budget for a stream segment.

The determination of complete budgets is rare. Omissions of important aspects, such as accurate measurements of streambed area, are frequent. Knowing the streambed area allows the investigator to express exports in terms of mass per streambed surface per unit time, facilitating comparisons among streams. However, very often (see Webster & Meyer 1997) CPOM exports are expressed as kg y^{-1}, making comparisons of streams differing in size difficult. Similarly, some areas of the stream channel may dry up during the annual hydrological cycle, and this fact is often ignored despite its importance for accurate budget calculations. Furthermore, the validity of the assumption that natural streams are in steady state depends on both the temporal and spatial scales considered (Cummins et al. 1983). For periods from days to a few weeks, changes in storage can be negligible, unless high variations in stream discharge occur (Minshall 1996). Although budgets should ideally consider entire streams (e.g. Fisher & Likens 1973) and be based on long-term data (Cummins et al. 1983), most studies have been restricted to short reaches and periods of one year or less (Webster & Meyer 1997).

Organic matter budgets quantify storage and fluxes in the ecosystem, allowing estimates of utilization efficiencies of the organic matter under different riparian vegetation and assessments of changes due to disturbances. Furthermore, it is beneficial to incorporate the considerable effort invested in measuring leaf breakdown rates during the last few decades into calculations of CPOM budgets (e.g. Pozo et al. 1997, Molinero & Pozo 2003), which is a useful summary of the information covered in the earlier chapters of this section.

Inputs (amount and timing), storage and outputs (transport and breakdown) depend on the type of material and, therefore, on the riparian vegetation. Some studies have demonstrated that forest disturbance modifies export and turnover of benthic CPOM (e.g. Webster et al. 1990, Pozo et al. 1997), and that these changes can impact stream communities (Graça et al. 2002). These and other studies suggest that, despite the difficulties of accurately measuring different components of organic matter budgets, and taking into account variation across spatial or temporal scales, CPOM budgets can be a useful indicator to assess the effects of human-induced disturbances on stream ecosystem functioning. This chapter describes procedures to construct CPOM budgets in streams. Table 7.1 shows selected CPOM turnover rates using data of terrestrial CPOM inputs and instream storage from the literature.

2. SITE SELECTION AND EQUIPMENT

2.1. Site Selection

The construction of a CPOM budget is easiest in narrow, forested headwater streams, although it can be performed in any wadeable stream. An entire small stream can be considered, but a 100-m stream segment will suffice.

2.2. Equipment and Material

Litter-fall inputs are determined from basket traps, randomly placed over the stream or in the riparian forest, provided the canopy cover in the channel and in the terrestrial environment is similar. Use lateral traps to determine lateral inputs of CPOM. Follow instructions in Chapter 1 for equipment, and materials. Use equipment, and materials presented in Chapter 4 for measuring CPOM storage.

If available, apply local leaf breakdown rates (see Chapter 6) to estimate losses by biological decomposition and physical processes. Otherwise, use suitable literature values for breakdown rates. Alternatively, use locally measured respiration rates of CPOM (see Chapter 31) or appropriate literature values.

J. POZO

*Table 7.1. Turnover rates (i.e. terrestrial inputs/storage) of coarse particulate organic matter
(excluding large wood) calculated from selected streams*

Location	Latitude	Vegetation	Stream order	Turnover (y^{-1})	Reference
Alaska, USA	65 °N	Taiga	1	13.8	1
			2	7.6	1
Denmark	56 °N	Beech forest	1	3.8	2
Quebec, Canada	50 °N	Spruce and deciduous forest	1	0.8	3
			2	0.9	3
			5	0.1	3
Oregon, USA	45 °N	Coniferous forest	1	0.3—0.7	4
			3	1.9	4
			5	12.0	4
New Hampshire, USA	44 °N	Deciduous forest	2	1.0	5
Spain	43 °N	Deciduous forest	1	16.2*	6
		Eucalyptus plantation	1	5.3*	6
Pennsylvania, USA	40 °N	Agricultural and wooded land	3	2.4	7
Virginia, USA	37 °N	Mixed forest	1	0.3	8
North Carolina, USA	35 °N	Deciduous and Rhododendron forest	1	3.9	9
		Deciduous forest	1	1.6	10
		Logged deciduous forest	2	1.5	10
Arizona, USA	33 °N	Desert scrub	5	3.8	11
South Africa	33 °S	Fynbos	2	7.6—13.9	12
Victoria, Australia	37 °S	Eucalyptus forest	4	6.0	13

*Leaves only. 1 = Irons & Oswood (1997); 2 = Iversen et al. (1982); 3 = Naiman & Link
(1997); 4 = Webster & Meyer (1997); 5 = Fisher & Likens (1973); 6 = Molinero & Pozo
(2003); 7 = Newbold et al. (1997); 8 = Smock (1990); 9 = Wallace et al. (1999); 10 = Webster
et al. (1990); 11 = Jones et al. (1997); 12 = King et al. (1987); 13 = Treadwell et al. (1997).

3. EXPERIMENTAL PROCEDURES

3.1. Sampling

1. Measure the stream segment length and, repeatedly, channel width along the
   entire reach to calculate the average and the stream segment surface area.
2. Measure widths and depths at the upper and lower ends of the stream reach to
   obtain the cross-sections profiles and determine discharge (see Chapter 1).
3. Terrestrial inputs: collect materials in litter-fall baskets and in lateral traps as
   described in Chapter 1.
4. Inputs and outputs by transport: collect material retained by drift nets (1-mm
   mesh size) located at both the upper and lower ends of the stream segment, and
   spanning the entire width of the channel as described in Chapter 1.

5. Storage: collect benthic CPOM as described in Chapter 4, taking care to sample entire channel transects (i.e. including dry parts). Beginning sampling downstream and move upstream to avoid extra CPOM transport caused by the disturbance of the stream bed.
6. Breakdown/respiration: If breakdown rates of leaves are measured, follow the instructions in Chapter 6 If respiration of the entire CPOM, or some of its components, is measured, follow the description in Chapter 31.

3.2. Laboratory Procedures

1. Separate CPOM from inorganic materials in all samples with the aid of nested sieves and under tap water when necessary (see Chapters 1 and 4).
2. Determine the dry mass and ash-free dry mass (AFDM) of the collected material (see Chapters 1 or 4).

3.3. Calculations

1. Calculate ash-free dry mass from laboratory determinations for total CPOM and its components as described in Chapters 1 and 4.
2. Express direct litter-fall inputs (I_D) in terms of g AFDM m^{-2} d^{-1} by dividing the amount of collected material by the surface area of the basket and by the number of days elapsed between successive sampling dates.
3. Express lateral inputs (I_L) in terms of g AFDM m^{-2} d^{-1} by dividing twice the amount of material collected in a given lateral trap ($2M_L$) by the length of the trap in meters (l), the mean channel width in meters (W) and the elapsed time in days (t):

$$I_L = \frac{2 \cdot M_L}{l \cdot W \cdot t} \qquad (7.3)$$

4. Alternatively, divide M_L by l, multiply by the total length of banks (2 × channel length) and divide by the area of the stream segment (channel length × mean channel width) and by the number of days (t).
5. Express CPOM transport (inputs, I_T, and outputs, O_T) in terms of g AFDM m^{-2} d^{-1} according to Chapter 1.
6. Calculate CPOM storage (S) in terms of g AFDM m^{-2} according to Chapter 4. To express change in storage in terms of g AFDM m^{-2} d^{-1}, calculate the difference in standing stock between sampling dates and divide by the number of days.
7. If breakdown rates (k) for particular leaf species are used, calculate them from leaf mass losses measured with litter bags according to Chapter 6, or use published breakdown rates of the dominant leaf species in the stream.
8. If respiration rates are used (the most frequent measures refer to benthic respiration measurements with no components differentiation; Webster et al.

1995), express them in terms of g AFDM m^{-2} d^{-1} according to Chapter 6. Alternatively, use published rates with the appropriate conversion factors.

9. Assuming that the stream is in steady state for inputs and outputs, calculate the turnover of the standing stock (T; d^{-1}) as the ratio of total litter input (I), in g m^{-2} d^{-1}, to the litter standing stock (S), in g m^{-2}:

$$T = \frac{I}{S} \qquad (7.4)$$

10. Litter inputs are either from the riparian vegetation (i.e. direct plus lateral inputs, $I_D + I_L$; g m^{-2} d^{-1}) or by transport from upstream (I_T; g m^{-2} d^{-1}). Similarly, litter outputs are either by transport downstream (O_T) or by breakdown (O_B). Thus, turnover can be also determined as the sum of the loss rates by transport (k_T; d^{-1}) and breakdown (k; d^{-1}). Thus:

$$\frac{I_D + I_L + I_T}{S} = \frac{O_T + O_B}{S} = k_T + k \qquad (7.5)$$

11. We can now estimate outputs by downstream transport (O_T; g m^{-2} d^{-1}) and by breakdown (O_B; g m^{-2} d^{-1}) as:

$$O_T = S \cdot k_T \qquad (7.6)$$

$$O_B = S \cdot k \qquad (7.7)$$

12. In the same way as the quotient respiration/inputs is an ecosystem efficiency (e.g. Fisher & Likens 1973, Webster & Meyer 1997), a measure of the stream ecosystem efficiency based on breakdown is proposed (Stream Breakdown Index, *SBI*) and can be obtained as:

$$SBI = \frac{O_B}{I} \qquad (7.8)$$

13. Other ratios (i.e. quotients between budget components) with ecological meaning can be calculated.

14. Compare whether calculated values of a given variable match the results of its measurement (i.e. validation).

4. FINAL REMARKS

A CPOM budget can be constructed regardless of the time interval considered (e.g. a week or a month). However, if an annual budget is to be constructed, samples should be collected over at least one year, twice a month during heavy litter fall and at least monthly during the rest of the year. In addition, special care needs to be taken to sample CPOM transport adequately during storms.

Budgets can be refined by sorting the collected CPOM into categories: leaves sorted into species, twigs (less than 1 cm in diameter), bark, fruits and flowers, and debris (unidentified fragments).

5. REFERENCES

Abelho, M. (2001). From litterfall to breakdown in streams: a review. *The ScientificWorld*, 1, 656-680.

Cummins, K.W., Sedell, J.R., Swanson, F.J., Minshall, Fisher, S.G., Cushing, C.E., Peterson, R.C. & Vannote, R.L. (1983). Organic matter budgets for stream ecosystems: problems in their evaluation. In: J.R. Barnes & G.W. Minshall (eds.), *Stream Ecology. Application and Testing of General Ecological Theory* (pp. 299-353). Plenum Press. New York.

Fisher, S.G. & Likens, G.E. (1973). Energy flow in Bear Brook, New Hampshire: an integrative approach to stream ecosystem metabolism. *Ecological Monographs,* 43, 421-439.

Graça, M.A.S., Pozo, J., Canhoto, C. & Elosegi, A. (2002). Effects of *Eucalyptus* plantations on detritus, decomposers, and detritivores in streams. *The ScientificWorld*, 2, 1173-1185.

Irons, J.G. & Oswood, M.W. (1997). Organic matter dynamics in 3 subarctic streams of interior Alaska, USA. In: J.R. Meyer & J.L. Meyer (eds.), Stream organic matter budgets. *Journal of the North American Benthological Society,* 16, 3-161, pp. 23-28.

Iversen, T., Thorup, J. & Skriver, J. (1982). Inputs and transformation of allochthonous particulate organic matter in a headwater stream. *Holarctic Ecology*, 5, 10-19.

Jones J.B., J.D. Schade, S.G. Fisher & N.B. Grimm. (1997). Organic matter dynamics in Sycamore Creek, a desert stream in Arizona, USA. In: J.R. Webster & J.L. Meyer (eds.), Stream organic matter budgets. *Journal of the North American Benthological Society*, 16, 3-161, pp. 23-28.

King, J.M., Day, J.A., Davies, B.R. & Hensall-Howard, M.P. (1987). Particulate organic matter in a mountain stream in the South-western Cape, South Africa. *Hydrobiologia*, 154, 165-187.

Larrañaga, S., Díez, J.R., Elosegi, A, & Pozo, J. (2003). Leaf retention in streams of the Agüera basin (northern Spain). *Aquatic Sciences*, 65, 158-166.

Minshall, G.W. (1996). Organic matter budgets. In: F.R. Hauer & G.A. Lamberty (eds.), *Methods in Stream Ecology* (pp. 591-605). Academic Press. San Diego.

Molinero, J. & Pozo, J. (2003). Balances de hojarasca en dos arroyos forestados: impacto de las plantaciones de eucalipto en el funcionamiento ecológico de un sistema lótico. *Limnetica*, 22, 65-73.

Naiman, R.J. & Link, G.L. (1997). Organic matter dynamics in 5 subarctic streams, Quebec, Canada. In: J.R. Webster & J.L. Meyer (eds.), Stream organic matter budgets. *Journal of the North American Benthological Society,* 16, 3-161. pp 33-39.

Newbold, J.D., Bott, T.L., Kaplan, L.A., Sweeney, B.W. & Vannote, R.L. (1997). Organic matter dynamics in White Clay Creek, Pennsylvania, USA. In: J.R. Webster & J.L. Meyer (eds.), Stream organic matter budgets. *Journal of the North American Benthological Society,* 16, 3-161. pp 46-50.

Pozo, J., González, E., Díez J.R. & Elosegi, A. (1997). Leaf-litter budgets in two contrasting forested streams. *Limnetica*, 13, 77-84.

Smock, L.A. (1990). Spatial and temporal variation in organic matter storage in low-gradient, headwater streams. *Archiv für Hydrobiologie*, 118, 169-184.

Treadwell, S.A., Campbell, I.C. & Edwards., R.T. (1997). Organic matter dynamics in Keppel Creek, southeastern Australia. In: J.R. Webster & J.L. Meyer (eds.), Stream organic matter budgets. *Journal of the North American Benthological Society*, 16, 3-161. pp 58-61.

Wallace, J.B., Eggert, S.L., Meyer, J.L. & Webster, J.R. (1999). Effects of resource limitation on a detrital-based ecosystem. *Ecological Monographs*, 69, 409-442.

Webster, J.R. & Meyer, J.L. (eds.). (1997). Stream organic matter budgets. *Journal of the North American Benthological Society*, 16, 3-161.

Webster, J.R., Golladay, S.W., Benfield, E.F., D'Angelo, D.J. & Peters, G.T. (1990). Effects of forest disturbance on particulate organic metter budgets of small streams. *Journal of the North American Benthological Society*, 9, 120-140.

Webster, J.R., Wallace, J.B. & Benfield, E.F. (1995). Organic processes in streams of the Eastern United States. In: C.E. Cushing, K.W. Cummins & G.W. Minshall (eds.), *Ecosystems of the World 22: River and Stream Ecosystems* (pp 117-187). Elsevier. Amsterdam.

PART 2

LEAF CHEMICAL AND PHYSICAL PROPERTIES

CHAPTER 8

DETERMINATION OF TOTAL NITROGEN AND PHOSPHORUS IN LEAF LITTER

MOGENS R. FLINDT[1] & ANA I. LILLEBØ[2]

[1]Biological Institute, SDU - University of Southern Denmark, Odense, Campusvej 55, 5230 Odense M, Denmark; [2]Departamento de Zoologia, Universidade de Coimbra, 3004-517 Coimbra, Portugal.

1. INTRODUCTION

In many freshwater and coastal marine environments plant litter is abundant and an important source of energy for the aquatic food webs. Nitrogen (N) and phosphorus (P) are important nutrients determining the quality of plant litter, and consequently their decomposition rates (Enriquez et al. 1993). Most phosphorus is used for the synthesis of ATP, RNA, DNA, and phospholipids (e.g. Sterner & Elser 2002), while nitrogen is mainly associated with proteins (see Chapter 9, Flindt et al. 1999). Generally, fast growing plants have low C:N and C:P ratios (high N and P content) and low fibre content, and decompose fast, whereas plants with slow growth rates exhibit slower litter decomposition rates (Enriquez et al. 1993, Flindt et al. 1999, Hill & Perrot 1995, Wrubleski et al. 1997). Table 8.1 summarizes nitrogen and phosphorus contents and C:N and C:P ratios of leaf litter of selected plant species.

A large fraction of the phosphorus is rapidly leached from dead leaf tissue, although P may be retained when autumn-shed leaves enter the aquatic environment before they dry out (Gessner 1991). Nitrogen is not generally leached upon senescence and death, although initially decreasing concentrations have been observed in some studies (Meyer & Johnson 1983). During decomposition, N and P concentrations of leaves usually increase (Webster & Benfield 1986, Gessner 1991). This increase is attributed to microbial colonization, which enhances the nutritional quality of leaf litter for detritivores (Webster & Benfield 1986).

The easiest way to quantify total N concentrations of leaves is with a CHN analyser. However, this equipment is expensive and not readily available in many laboratories. The method described here is a modification of the standard Kjeldahl-N method for plant samples (Ferskvandsbiologisk Laboratorium University of Copenhagen 1992). Samples of plant material are dried, homogenized and digested

M.A.S. Graça, F. Bärlocher & M.O. Gessner (eds.), Methods to Study Litter Decomposition:
A Practical Guide, 53 – 60.

in concentrated sulphuric acid to reduce all nitrogen species to ammonia (NH_3-N). After neutralizing the resulting solutions, they are filtered and analyzed spectrophotometrically like water samples.

The method described for P involves drying, homogenization, and combustion of leaf litter samples. The plant ashes are digested in concentrated hydrochloric acid. The samples are then filtered and total P is quantified as phosphate (PO_4-P) using the ascorbic acid method.

A combined method for total N and total P determination can be performed by following the total N procedure, but the pH during neutralization of samples should never exceed 6 in this procedure, otherwise dissolved phosphorus will precipitate. The samples are then filtered and analysed like water samples.

Ebina et al. (1983), Ostrofsky (1997) and Gessner et al. (1998) are examples of other methods that can be used to determine total N and/or total P concentrations in leaf litter.

Table 8.1. Nitrogen and phosphorus concentrations of leaf litter of selected plants.

Leaf material	N (%)	C:N	P (%)	C:P	Reference
2 macroalgae	2.50—3.92	12.6—18.7	0.19—0.36	224—544	1
2 seagrass species	2.64—2.76	15.6—24.3	0.50—0.55	191—258	1
4 freshwater angiosperms	1.37—3.44	12.7—27.8	0.43—0.85	140—249	1
6 sedge species	0.18—1.07	46.1—315	0.016—0.150	729—7847	1
2 pine species	0.40—1.51		0.017—0.131		1
6 oak species	0.53—1.03	46.3—93.6	0.080—0.291	167—596	2
4 birch species	0.70—2.30	20.7—67.9	0.065—0.381	116—731	2
5 maple species	0.73—1.71	24.1—60.9	0.112—0.411	108—368	2
2 willow species	0.83—2.24	21.0—56.5	0.121—0.281	167—389	2
3 poplar species	0.92—2.38	19.7—52.4	0.083—0.092	533—581	2
5 evergreen shrub and tree species	~1.4—2.3	~10—19			3
14 deciduous shrub and tree species	~2.4—3.6	~20—30			3

1 = Enriquez et al. (1993); 2 = Ostrofsky (1997); 3 = Pérez-Harguindeguy et al. (2000).

2. NITROGEN: EQUIPMENT, CHEMICALS AND SOLUTIONS

2.1. Equipment and Material

- Drying oven
- Mill or mortar and pestle
- Spectrophotometer
- Hot plate
- pH meter

- Analytical balance (± 0.1 mg)
- Vortex
- Glass fibre filters (e.g. GF/C, Whatman)
- Syringes (20 ml)
- Pipettes (0.2, 1 and 5 ml)
- Erlenmeyer flasks (50 ml)
- Glass tube
- Cooling bulbs

2.2. Chemicals (analytical grade)

- Deionized water
- $CuSO_4 \cdot 5H_2O$
- H_2SO_4 (98%)
- $(NH_4)SO_4$
- Thymol (crystalline)
- Phenol (crystalline)
- $Na_2[Fe(CN)_5NO] \cdot 2H_2O$ (sodium nitroprusside dihydrate)
- NaOH
- NaOCl (15%)
- Sodium citrate dihydrate

2.3. Solutions

- Solution 1: 10 g $CuSO_4 \cdot 5H_2O$ per 100 ml of deionized water.
- Solution 2: NH_3-N stock solution (100 mg N l^{-1}). Dissolve 0.472 g $(NH_4)_2SO_4$ in 1000 ml of deionized water. Add one crystal of thymol. The stock solution can be kept for three months at ambient temperature.
- Reagent A: 50 g phenol and 0.25 g sodium nitroprusside dihydrate in 1000 ml of deionized water.
- Reagent B: 25 g NaOH and 20 ml 15% NaOCl in 1000 ml of deionized water. If plant samples are from estuarine or marine environments, add 20 g of sodium citrate dihydrate per 100 ml, to avoid high turbidity.

3. NITROGEN: EXPERIMENTAL PROCEDURES

3.1. Sample Preparation

1. Dry the plant material at 40—50 °C, grind it and pass the powder through a 0.5 mm mesh screen.
2. Weigh portions of about 5 mg to the nearest 0.1 mg; correct for water content.
3. Place the sample in a 50 ml Erlenmeyer flask with 25 ml of deionized water.
4. Add 0.2 ml of Solution 1 and 1 ml H_2SO_4.

5. Place the Erlenmeyer flasks on a hot plate and boil off water.
6. After the water has evaporated a light smoke will appear. At this point, a cooling bulb should be placed on top of the flasks.
7. The digestion of the organic matter starts, when a dark coloured tar appears. Rotate the flask from time to time to remove residues from the walls. When the digestion is complete, the acid solution becomes light green and transparent. Continue the heating for another 30 min.
8. After cooling, transfer the solution to a glass and rinse the flask walls with deionised water. Adjust the samples with NaOH to the point where Cu $(OH)_2$ flocculates. The pH should be between 5 and 8.
9. Adjust the sample volume to 100 ml with deionised water.
10. Filter the samples through a GF/C filter connected to a 20 ml syringe.

3.2. Spectrophotometric Analysis

1. Transfer 10 ml of sample to a glass tube.
2. Add 1.0 ml of Reagent A to 10 ml of sample and vortex.
3. Add 1.0 ml of Reagent B and mix again. Shield the developing blue colour from sunlight.
4. Wait at least 1 h for the reaction to complete.
5. Measure absorbance at 630 nm, using deionized water as reference.
6. Use 10 ml of deionized water, treated like the samples, as a blank.
7. Run a standard curve with the following concentrations: 100, 200, 400, 800 µg NH_3-N l^{-1}. The standard curve is linear up to 1000 µg NH_3-N l^{-1}.
8. Calculate the N concentration in plant litter based on the standard curve taking into account any dilutions and the water content of the ground plant material.

4. PHOSPHORUS: EQUIPMENT, CHEMICALS AND SOLUTIONS

4.1. Equipment and Material

- Drying oven
- Mill or mortar and pestle
- Muffle furnace
- Spectrophotometer
- Hot plate
- Analytical balance (± 0.1 mg)
- Vortex
- Glass fibre filters (e.g. GF/C, Whatman)
- Syringes (20 ml), acid-washed
- Pipettes (0.2, 1 and 5 ml)
- Erlenmeyer flasks (50 ml), acid-washed
- Glass tube, acid-washed
- Cooling bulbs, acid-washed

4.2. Chemicals (analytical grade)

- Deionized water
- HCl (37%)
- KH_2PO_4
- Thymol (crystalline)
- $(NH_4)_6Mo_7O_{24} \cdot 4H_2O$ (ammonium heptamolybdate tetrahydrate)
- H_2SO_4 (98%)
- $K(SbO)C_4H_4O_6 \cdot 5H_2O$ (potassium antimony (III) oxide tartrate pentahydrate, extra pure)
- L(+)ascorbic acid (vitamin C)

4.3. Solutions

- Solution 1: PO_4-P stock solution (40 mg P l^{-1}). Dissolve 175.75 mg KH_2PO_4 in 1000 ml of deionized water. Add one crystal of thymol. The solution can be kept for three months at ambient temperature.
- Solution 2: Stock solution of reagent. Dissolve 12 g of $(NH_4)6Mo_7O_{24} \cdot 4H_2O$ into 500 ml of deionized water; add very carefully 140 ml of concentrated H_2SO_4. After mixing and cooling, add 275 mg of $K(SbO)C_4H_4O_6 \cdot 5H_2O$ and let it dissolve; adjust the volume to 1000 ml with deionized water. The solution can be kept for three months at ambient temperature.
- Solution 3: Working solution of reagent. Immediately before use add 1.06 g of ascorbic acid to 100 ml of stock solution of the reagent (Solution 2).

5. PHOSPHORUS: EXPERIMENTAL PROCEDURES

5.1. Sample Preparation

1. Dry plant material, grind it and pass the powder through a 0.5 mm mesh screen.
2. Place dried powder in the muffle furnace for 4 h at 500 °C to combust the organic compounds.
3. Weigh out 5 mg of ash.
4. Place the sample in a 50 ml Erlenmeyer flask with 25 ml of deionized water.
5. Add 1 ml of conc. HCl.
6. Place the Erlenmeyer flask on a heating plate to evaporate the water; place a cooling bulb on top of the flasks.
7. Continue until the solution turns yellow and transparent. Rotate the flask from time to time to remove ash particles from the walls.
8. Adjust sample volume to 100 ml with deionized water.
9. Filter the sample through a GF/C filter connected to a 20 ml syringe.

5.2. Spectrophotometric Analysis

1. Place 10 ml of the sample in a glass tube.
2. Add 1.0 ml of the reagent and vortex.
3. After waiting for at least 15 min, measure absorbance at 882 nm, using deionized water as reference.
4. Use 10 ml of deionized water, treated like the sample, as blank.
5. Run a standard curve using the following concentrations: 40, 80, 100, 200, 400, 800 µg PO_4-P l^{-1}. The standard curve is linear up to 1000 µg PO_4-P l^{-1}.
6. Calculate the P concentration in plant litter based on the standard curve and taking into account any dilutions and the water content of the ground plant material.

6. REFERENCES

Ebina, J., Tsutsui, T. & Shirai, T. (1983). Simultaneous determination of total nitrogen and total phosphorus in water using peroxodisulfate oxidation. *Water Research,* 17, 1721-1726.

Enríquez, S., Duarte, M., & Sand-Jensen, K. (1993). Patterns in decomposition rates among photosynthetic organisms: the importance of detritus C:N:P content. *Oecologia*, 94, 457-471.

Flindt, M.R., Pardal, M.A., Lillebø, A.I., Martins, I. & Marques, J. C. (1999). Nutrient cycling and plant dynamics in estuaries: A brief review. *Acta Oecologica*, 20, 237-248.

Gessner, M.O. (1991). Differences in processing dynamics of fresh and dried leaf litter in a stream ecosystem. *Freshwater Biology*, 26, 387-398.

Gessner, M.O., Robinson C.T. & Ward, J.V. (1998) Leaf breakdown in streams of an alpine glacial floodplain: dynamics of fungi and nutrients. *Journal of the North American Benthological Society*, 17, 403-419.

Hill, B.H. & Perrot Jr., W.T. (1995). Microbial colonisation, respiration, and breakdown of maple leaves along a stream-marsh continuum. *Hydrobiologia*, 312, 11-16.

Ferskvandsbiologisk Laboratorium University of Copenhagen ed. (1992) Limnologisk Metodik. Akademisk Forlag. Copenhagen. (Nitrogen method pp. 28-29 and 36-37; Phosphorus method pp. 40-41 and, 106-107).

Meyer, J.L. & Johnson, C. (1983). The influence of elevated nitrate concentration on rate of leaf decomposition in a stream. *Freshwater Biology*, 13, 177-184.

Ostrofsky, M.L. (1997). Relationship between chemical characteristics of autumn-shed leaves and aquatic processing rates. *Journal of the North American Benthological Society*, 16, 750-759.

Pérez-Harguindeguy, N., Díaz, S., Cornelissen, J.H.C., Vendramini, F., Cabido, M. & Castellanos, A. (2000). Chemistry and toughness predict leaf litter decomposition rates over a wide spectrum of functional types and taxa in central Argentina. *Plant and Soil*, 218, 21-30.

Sterner, R.W. & Elser, J.J. (2002). *Ecological Stoichiometry: The Biology of Elements from Molecules to the Biosphere*. Princeton University Press. Princeton.

Webster, J.R. & Benfield, E.F. (1986). Vascular plant breakdown in freshwater systems. *Annual Review of Ecology and Systematics*, 17, 567-594.

Wrubleski, D.A., Murkin, H.R., van der Valk, A.G. & Nelson, J.W. (1997). Decomposition of emergent macrophyte roots and rhizomes in the northern prairie marsh. *Aquatic Botany*, 58, 121-134.

CHAPTER 9

TOTAL PROTEIN

MARK O. BAERLOCHER

*Medical Education Office, St Michael's Hospital, 30 Bond Street, School of Medicine,
University of Toronto, Toronto, ON, Canada, M5B 1W8.*

1. INTRODUCTION

Detrital plant matter, which supplies most of the energy and nutritional needs for stream communities (Allan 1995), consists primarily of structural plant polysaccharides (Chapters 11 and 17). These are generally inaccessible to invertebrates without microbial assistance, i.e., conditioning (Chapters 25 and 30). An important aspect of microbial conditioning is the enrichment of the substrate with nitrogen, especially in the form of protein, derived from microbial cells and excretions (Kaushik & Hynes 1971, Bärlocher et al. 1989). During decomposition, nitrogen and protein levels tend to decline somewhat in the first few days and then often rise to a level that may remain constant for several weeks (Fig. 9.1).

Factors that influence protein concentrations in decomposing leaves include leaf species, concentrations of dissolved nutrients in the stream (primarily N and P), and length of exposure in the stream. Estimated values can also depend critically on the analytical method that is used. A common approach in forage analysis is to multiply the nitrogen concentration (normally estimated by the Kjeldahl procedure) with an empirical factor of 6.25 (AOAC 1990). When applied to leaves decomposing in streams, this gives protein concentrations higher than those estimated with other assays (Kaushik & Hynes 1971, Gessner 1991).

Nitrogen is not restricted to protein; it occurs, for example, in chitin (a constituent of fungal cell walls) or in 'artifact lignins', complexes that may form between leaf phenolics and compounds released by microbes during decomposition (Suberkropp et al. 1976, Odum et al. 1979). An alternative to deriving protein concentrations from analyses of N is to hydrolyze any proteins present in a sample to amino acids which can then be quantified by HPLC (Craig et al. 1989, Chapter 10) or other assays. Similarly, to estimate amounts of protein available (digestible) to detritivores, the sample can be exposed to invertebrate enzyme extracts and the released amino acids measured (Bärlocher 1983, Bärlocher & Howatt 1986, Craig et

*M.A.S. Graça, F. Bärlocher & M.O. Gessner (eds.), Methods to Study Litter Decomposition:
A Practical Guide, 61 – 68.*

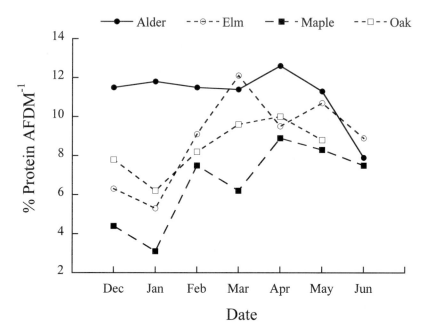

Fig. 9.1. Protein contents (as % of ash-free dry mass) of leaves decaying in the Speed River (Kaushik & Hynes 1971). The leaves were exposed on December 7, and recovered at monthly intervals. Alder: Alnus glutinosa; Elm: Ulmus americana; Maple: Acer saccharum; Oak: Quercus alba.

al. 1989). These methods are relatively elaborate and not needed where the objective is simply to estimate total protein concentrations.

The first step in protein analysis is extraction from the substrate (Scopes 1982). Protein solubility is pH dependent: in the presence of phenolics, complexes tend to form that are most stable near neutrality and dissociate more readily with increasing alkalinity (Swain 1979) or in the presence of surfactants (Bärlocher et al. 1989). The protein-containing extract is then mixed with a chemical that reacts stoichiometrically to form a coloured product measured in the visible region of the spectrum. For example, the Biuret method is based on the formation of a violet complex between cupric (Cu^+) ions and two or more peptide bonds. The intensity of the colour is proportional to the number of peptide bonds (Bailey 1962). In its original form it is relatively insensitive. Lowry et al. (1951) combined the copper reaction in Biuret's approach with Folin's reagent to produce a strong dark blue colour with the phenolic amino acid tyrosine found in proteins (cf. Chapter 14).

Unfortunately, many substances interfere with this assay, not least among them non-proteinaceous phenolics that may be present in high concentrations in leaf litter. To avoid this problem, proteins can be precipitated from the extract with TCA (trichloroacetic acid), which does not react with other phenolics. The pellets can then be redissolved in 0.1 N NaOH, and the reaction is performed with this new solution (Kaushik & Hynes 1971).

Another very popular and more sensitive method is based on the non-specific binding of Coomassie Blue (Brilliant Blue) to almost all proteins in an approximately stoichiometric manner (Bradford 1976). The binding results in a shift in the absorption maximum of the dye from 465 to 595 nm. This assay is generally less susceptible to interference by other substances than the Lowry or Biuret method, but strongly alkaline conditions and high concentrations of detergents can distort the result.

The results from samples are usually compared to a calibration curve with a defined protein. For greatest accuracy, the calibration protein should be one that dominates in the sample. This has rarely been attempted in ecological studies. For convenience, bovine serum albumin (BSA) has most often been used. Since leaf litter contains a mix of proteins, any estimates of total protein based on a single standard are approximations.

This chapter introduces the Bradford assay, based on the binding of Coomassie Blue to protein, to follow changes in protein concentrations during leaf senescence and decomposition. The protocol has been adopted from Baerlocher et al. (2004). It involves extraction of total protein (i.e., both hydrophobic and hydrophilic proteins) from both leaf material and microorganisms such as fungi, should the leaf be colonized.

2. EQUIPMENT, CHEMICALS AND SOLUTIONS

2.1. Equipment and Material

- Leaves of various stages
- Thermos to carry liquid nitrogen (optional)
- Freezer at –80 °C (optional)
- Analytical balance
- Drying oven
- Mortar and pestle
- Sterile sand
- Microcentrifuge, capable of 12,000 g
- Eppendorf centrifuge tubes (1.5 ml)
- Ultrasonication probe (e.g. Sonic 300 Dismembrator, Artek Systems, 300 W)
- Spectrophotometer
- Cuvettes (1 cm)
- Liquid nitrogen (optional)
- Crushed ice
- Bowl
- Hotplate and boiling water bath
- Parafilm

- Brown bottle
- Test tubes or small bottles with capacity of at least 3 ml
- 200 µl pipettor
- Graduated cylinder, or 1000 µl pipettor
- Scanner (optional)
- Glass fiber filters (optional)

2.2. Chemicals (analytical grade)

- Trizma® base
- Sucrose
- Ethlylene diamine tetraacetic acid (EDTA)
- Sodium dodecyl sulfate (SDS)
- Dithiothreitol (DTT)
- Double distilled water (ddH$_2$O)
- Coomassie Blue G
- Methanol
- H$_3$PO$_4$ (85%)
- Bovine serum albumin (BSA)

2.3. Solutions

- Solubilization buffer: dissolve 100 nM Trizma base, 160 nM sucrose, 1 nM EDTA, 1% SDS (w:v) in ddH$_2$O; store at 4 °C. Add 0.5% SDS immediately before use.
- Dye stock solution: dissolve 100 mg of Coomassie Blue G in 50 ml methanol. If the solution is turbid, filter through a glass fiber filter. Add this solution to 100 ml of 85% H$_3$PO$_4$, and then top up to 200 ml with ddH$_2$O. Final concentrations of the dye stock are therefore 0.5 mg ml^{-1} Coomassie Blue G, 42.5% H$_3$PO$_4$, and 25% methanol. The dye stock should appear dark red, and have a pH close to 0. The dye stock solution may be stored indefinitely in a brown bottle at 4 °C.
- Assay reagent: prepare the assay reagent by diluting the dye stock solution 1:4 with ddH$_2$O. The solution should appear brown, and have a pH of 1.1. This diluted solution is stable in a brown bottle at 4 °C for weeks.

3. EXPERIMENTAL PROCEDURES

3.1. Leaf Protein Extraction

1. Collect leaves in the field. With senescent leaves, it is best to place them immediately in liquid nitrogen and in the laboratory, store at –80 °C to minimize protease activity.

2. When ready for extraction, a sample is gently cleaned and dried with paper towels.
3. With scissors or a cork borer, cut out between 0.5 and 2 cm^2 of leaf tissue, and determine its fresh mass to the nearest 0.1 or 0.01 mg. The sample is flattened under a glass plate in a scanner, and scanned to determine leaf area.
4. Dry at 40—50 °C to constant mass and reweigh to determine dry mass.
5. Grind each sample for ca. 10 min into a fine powder using a clean mortar and pestle, liquid nitrogen (optional) and 50 mg sand.
6. If living or senescent leaves are used, the samples should not be allowed to thaw at any time from this point on, as this may activate proteases. Additional liquid nitrogen must therefore periodically be added to the sample as it is being ground. If the sample strongly adheres to the pestle, a longer drying period may be required.
7. Transfer powdered sample to a 1.5 ml Eppendorf centrifuge tube.
8. Clean mortar and pestle by adding another 50 mg of sand, grind for an additional 2 min and transfer powder to the same Eppendorf tube. Repeat a third time. Periodically add more liquid nitrogen. Keep the Eppendorf tubes on ice at all times.
9. Add 250 µl of solubilization buffer to each Eppendorf microcentrifuge tube and vortex.
10. Add liquid nitrogen to a mortar, then dip the Eppendorf microcentrifuge tube into the nitrogen freezing the contents. Sonicate the tube until the contents have thawed.
11. Repeat the freeze/thaw step.
12. Incubate tubes in a water bath at 90 °C for 5 min. Be careful that the tubes do not burst open, spraying some of their contents. Boil-proof tubes are available. Alternatively, secure the tube caps with Parafilm or labelling tape. Centrifuge tubes for 5 min at 12,000 g.
13. Apportion the supernatant with the protein into 2 tubes, and store in a freezer (preferably –80 °C) until use.

3.2. Calibration Curve

1. In small bottles or test tubes, prepare 6 protein standard solutions by dissolving bovine serum albumin (BSA) in the same buffer used for the experimental protein extract. Use the following final concentrations: 0, 250, 500, 1000, 1500 and 2000 µg ml^{-1}.
2. Set the spectrophotometer to a wavelength of 595 nm, over a range of 0 to 2 Absorbance Units (AU). Generally, the spectrophotometer should be allowed to warm up for at least 15 min before measurements are taken.
3. Set the spectrophotometer to zero when reading the absorbance of ddH$_2$O. Double-check that the spectrophotometer has been properly adjusted by taking a sample measurement of ddH$_2$O and ensuring it reads 0.

4. Into a clean cuvette, add 40 µl of the first protein standard to 2 ml assay reagent, cover cuvette with Parafilm, invert several times, and wait 15—60 min.
5. Record absorbance (optical density, O.D.) at 595 nm. Repeat this for each protein standard, in triplicate, to ensure accurate results.
6. Average the triplicate measurements for each protein standard concentration, and plot absorbance measured at 595 nm versus BSA concentration. Take into account the dilutions required in the procedure when plotting BSA concentration. For example, the BSA standard tube 2 initially had a BSA concentration of 250 µg ml^{-1}. This was diluted by a factor of 51 (40 µl standard was added to 2 ml assay reagent, giving a final total volume of 2.04 ml). Therefore, the BSA concentration to plot for tube 2 is 4.90 µg ml^{-1} (250 µg ml^{-1} × 0.04 ml / 2.04 ml).

3.3. Spectrophotometric Assay of Sample Extracts

1. To measure protein concentration of the sample extract, add 40 µl of the extract to 2 ml of assay reagent in a clean cuvette.
2. Cover cuvette with Parafilm, invert several times, and wait 15—60 min. Record absorbance at 595 nm.
3. Repeat this in triplicate for each extract.
4. If the absorbance measurement for any sample is greater than the 0—2 scale, repeat the reading after first diluting the extract, until the absorbance measurements fall inside the range.
5. Average the triplicate measurements from each sample, and estimate the protein concentration of the extract using the standard curve created earlier.
6. From the final, diluted solution in the cuvette, calculate the concentration of the initial, undiluted protein extract. For example, if the protein concentration of the sample extract is estimated to be 10 mg ml^{-1}, and the initial extract was diluted once by adding 40 µl to 2 ml of reagent (i.e. a dilution of 51 times), the protein concentration in the initial extract was 510 µg ml^{-1}. This value can then be converted into a value representing mg protein per g fresh or dry leaf mass, or per cm^2 of leaf area.

4. FINAL REMARKS

The Bradford assay only gives a linear standard curve within a relatively narrow range. For highest accuracy and precision, sample concentrations should be adjusted to fall within that range.

The protein extraction procedure described here breaks disulfide bonds. This allows separating and analyzing the extracted mix of proteins based on their sizes without additional treatment. For example, the mix can be analyzed by SDS-PAGE (sodium dodecylsulfide polyacrylamide gel electrophoresis).

4. REFERENCES

Allan, J.D. (1995). *Stream Ecology. Structure and Function of Running Waters*. Chapman & Hall. London.

AOAC (1990). Protein (crude) determination of animal feed: copper catalyst Kjeldahl method (984.13). *Official Methods of Analysis of the Association of Official Analytical Chemists*, 15th edition. AOAC International. Gaithersburg, Maryland.

Baerlocher, M.O., Campbell, D.A. & Ireland, R.J. (2004). Developmental progression of photosystem II electron transport and CO_2 uptake in *Spartina alterniflora*, a facultative halophyte, in a northern salt marsh. *Canadian Journal of Botany,* 82, 365-375.

Bailey, J.L. (1962). *Techniques in Protein Chemistry*. Elsevier. New York.

Bärlocher, F. (1983). Seasonal availability and digestibility of CPOM in a stream. *Ecology*, 64, 1266-1272.

Bärlocher, F. & Howatt, S.L. (1986). Digestion of carbohydrate and protein by *Gammarus mucronatus* Say (Amphipoda). *Journal of Experimental Marine Biology and Ecology*, 104, 229-237.

Bärlocher, F., Tibbo, P.G. & Christie, S.H. (1989). Formation of phenol-protein complexes and their use by two stream invertebrates. *Hydrobiologia*, 173, 243-249.

Bradford, M.M. (1976). A rapid and sensitive method for the quantitation of microgram quantities of protein using the method of protein dye bonding. *Analytical Biochemistry*, 72 248-254.

Craig, D., Ireland, R.J. & Bärlocher, F. (1989). Seasonal variation in the organic composition of seafoam. *Journal of Experimental Marine Biology and Ecology*, 130, 71-80.

Gessner, M.O. (1991). Differences in processing dynamics of fresh and dried leaf litter in a stream ecosystem. *Freshwater Biology*, 26, 387-398.

Kaushik, N.K. & Hynes, H.B.N. (1971). The fate of the dead leaves that fall into streams. *Archiv für Hydrobiologie*, 68, 465-515.

Lowry, O.H., Rosebrough, N.J., Farr, A.L. & Randall, R.J. (1951). Protein measurement with the Folin phenol reagent. *Journal of Biological Chemistry*, 193, 265-275.

Odum, W.E., Kirk, P.W. & Zieman, J.C. (1979). Non-protein nitrogen compounds associated with particles of vascular plant detritus. *Oikos*, 32, 363-367.

Scopes R. (1982). *Protein Purification*. Springer-Verlag. New York.

Suberkropp, K., Godshalk, G.L. & Klug, M.J. (1976). Changes in the chemical composition of leaves during processing in a woodland stream. *Ecology*, 57, 720-727.

Swain, T. (1979). Tannins and lignins. In: G.A. Rosenthal & D.H. Janzen (eds.), *Herbivores: Their Interactions with Secondary Plant Metabolites* (pp. 657-682). Academic Press. New York.

CHAPTER 10

FREE AMINO ACIDS

SHAWN D. MANSFIELD[1] & MARK O. BAERLOCHER[2]

[1]*Canada Research Chair in Wood and Fibre Quality, Department of Wood Science, University of British Columbia, Vancouver, B.C., Canada, V6T 1Z4;* [2]*Medical Education Office, St. Michael's Hospital, 30 Bond Street, School of Medicine, University of Toronto, Toronto, ON, Canada, M5B 1W8.*

1. INTRODUCTION

Amino acids are one of the components that may contribute to the leaching effect most commonly observed in dried leaves (Gessner 1991). Amino acids, along with soluble carbohydrates (Chapter 12) represent a readily digestible source of nutrients for most organisms. This is in contrast to low-molecular weight phenolics, which are also liberated during leaching, but are not easily metabolized by invertebrates or microorganisms.

The total concentration of free amino acids in decaying leaves represents a dynamic equilibrium between losses (leaching) and uptake (abiotic and biotic). Positively charged amino acids are generally adsorbed more readily to surfaces (Armstrong & Bärlocher 1989a), and are influenced by the concentration of Ca^{2+} in the stream water, which selectively affects the adsorption of various types of amino acids (Armstrong & Bärlocher 1989b). Uptake of amino acids by aquatic hyphomycetes from the stream water may affect the protein level of decaying leaves, as well as the fungal species that successfully colonize newly immersed leaves (Bengtsson 1988).

There are many approaches to identify and quantify amino acids in sample extracts. A quick and simple estimate of total amino acid concentration is based on the reaction of reduced ninhydrin with the amino group (Rosen 1957). A more accurate technique uses pre-column derivatization and high performance liquid chromatography (HPLC), to facilitate the identification of individual amino acids (Craig et al. 1989, Cohen 1990, Hodisan et al. 1998). The derivatized samples can be separated and quantified by fluorescence (Hodisan et al. 1998) or UV detection (Cohen 1990).

The details of the HPLC protocol used depend on the type and availability of HPLC systems. The following protocol should serve as a general guide, and may

M.A.S. Graça, F. Bärlocher & M.O. Gessner (eds.), Methods to Study Litter Decomposition: A Practical Guide, 69 – 74.

have to be modified. It should also be noted that polypeptides and proteins can be hydrolyzed into their individual amino acids (boiling in 6 N HCl under vacuum for 20—24 h; Craig et al. 1989). This protocol results in the conversion of tryptophan, glutamine and asparagine to their respective acids, and these amino acids are therefore not measured. However, the hydrolysis step is not included in the current protocol, as the intention here is to determine the identity and quantity of free amino acids (amino acids present in peptides or proteins are labelled 'bound amino acids'). We therefore include a step which removes polypeptides and proteins to prevent contamination of the HPLC column or co-elution. The presented method has been adopted from Hodisan et al. (1998).

2. EQUIPMENT, CHEMICALS AND SOLUTIONS

2.1. Equipment and Material

- Analytical balance
- Desiccator (containing phosphorus pentoxide)
- Micropipettors
- Scissors or cork borer
- Vortex
- Drying oven
- Mortar and pestle
- Sterile sand
- Centrifuge, capable of 12,000 g
- Thermos to carry liquid nitrogen (optional)
- Liquid nitrogen (optional)
- Freezer at –80 °C (optional)
- Scanner (optional)
- Glass fiber filters (optional)
- HPLC system with the capacity to run a gradient elution profile equipped with a UV detector
- C-18 reversed-phase column, i.e. Lichrospher ODS 100 (5 μm particle diameter; 250 × 4.6 mm)
- C-18 guard column
- pH meter
- HPLC vials and caps
- 0.45 μm HPLC filters
- Glassware such as test tubes, centrifuge tubes, graduated cylinders, acid-washed (e.g. soaked in 10% nitric acid overnight and then rinsed thoroughly with deionized distilled water).
- Organic matter samples (i.e. leaves harvested from trees or recovered from a stream).

2.2. Chemicals and Solutions

- Double distilled water (ddH$_2$O)
- HCl (1%)
- Acetonitrile (HPLC grade)
- 4-nitrophenylisothiocyanate in acetonitrile (NPITC), 50 mM stock solution; this must be prepared daily, and stored in a dark bottle at 4 °C.
- 10% triethylamine (TEA) in acetonitrile
- Derivatization reagent (working solution): 490 μl of the NPITC stock solution, 50 μl of the 10% TEA stock solution, and 50 μl of ddH$_2$O water.
- Toluene (HPLC grade)
- Glacial acetic acid
- Sodium acetate (0.14 M)
- Amino acid standards: Though it is possible to purchase individual amino acids to be used as standards, it is generally more efficient to use a commercially available standard mix with known amounts of several amino acids. Dilute the mix in ddH$_2$O to a known concentration of between 50 and 250 μM (100 μM is suggested).

3. EXPERIMENTAL PROCEDURE

3.1. Extraction of Free Amino Acids

1. Collect leaves in the field. When using living or senescent leaves, place them immediately in liquid nitrogen and in the laboratory store them at –80 °C to minimize protease activity.
2. When ready for extraction, a sample is gently cleaned and dried with paper towels.
3. With scissors or a cork borer, cut out between 0.5 and 2 cm^2 of leaf tissue and determine its fresh mass to the nearest 0.1 or 0.01 mg. The sample is flattened under a glass plate in a scanner, and scanned to determine leaf area.
4. Dry at 40—50 °C to constant weight and reweigh to determine dry mass.
5. Grind each sample for ca. 10 min into a fine powder using a clean mortar and pestle, liquid nitrogen (optional) and 50 mg sand.
6. If living or senescent leaves are used, the samples should not be allowed to thaw at any time from this point on, as this may activate proteases. Additional liquid nitrogen must therefore periodically be added to the sample as it is being ground. If the sample strongly adheres to the pestle, a longer drying period may be required.
7. Transfer powdered sample to a 10 ml centrifuge tube.
8. Add 5 ml of 0.1 % HCl and vortex.
9. Centrifuge tube for 5 min at 12,000 *g*.
10. The supernatant is unlikely to contain large amounts of proteins. Nevertheless, a guard column on the HPLC may be useful t to prevent clogging up the column.

11. Store the supernatant in a freezer (preferably at –80 °C) until use.

3.2. Derivatization of Amino Acids

1. Samples and standards are freeze-dried and stored overnight in a desiccator (containing phosphorus pentoxide) before derivatization.
2. Record sample weight from freeze-dried sample stock to the nearest 0.01 mg.
3. Derivatize dried samples and standards in Pyrex tubes by adding 25 µl of the derivatization working solution.
4. Vortex for 1 min and allow to react for 10 min.
5. After exactly 10 min add 100 µl of ddH$_2$O.
6. Extract reaction mixture with 125 µl of toluene with gentle mixing. Allow phases to separate, collect aqueous layer, and discard organic layer in appropriate container.
7. Filter aqueous sample through 0.45 µm filter into an HPLC vial and cap.

3.3. HPLC Analysis and Amino Acid Determination

1. Allow eluants (i.e. solvents used as mobile phase) to degas appropriately before HPLC analysis.
2. Set HPLC to the conditions indicated in Table 10.1.
3. Run standards on HPLC.
4. Determine retention time and peak area for each amino acid in the standard solution. Correcting for any dilutions (if required), use this information to determine the concentration of each individual amino acid in the standard.

Table 10.1. HPLC conditions for aminoacids determination.

Parameter	Condition
Eluent A	94% 0.14 M sodium-acetate, 0.05% TEA (v:v) in 6% acetonitrile, pH adjusted to 6.4 with glacial acetic acid
Eluent B	60% acetonitrile in deionized distilled dd H$_2$O
Column equilibration before injection	5 min with 85% eluant A and 15% eluant B
Gradient	Linear gradient from 85% eluant A and 15% eluant B to 40% eluant A and 60% eluant B over 30 min; wash column with 100% eluant B for 5 min, and re-equilibrate column for 10 min at 85% elute A and 15% eluant B.
Flow rate	1 ml min^{-1}
Column temperature	45 °C
Detection	UV detection at 340 nm (alternatively use 254 nm)
Injection volume	20 µl

5. Prepare standard curves by plotting areas under the peaks versus concentrations of each amino acid in the standards mixture.
6. Run each sample, always using the same volumes as for the calibration. If the sample peaks are off the scale, dilute with ddH$_2$O water.
7. Determine which peak represents which amino acid (by retention time), and subsequently determine sample concentrations (μmol μl^{-1}) based on the standard curves for each amino acid, using the linear equation ($y = mx + b$).
8. Adjust the total concentration of the individual amino acids in the sample for the total sample volume and divide by the total weight of material used (mg). The individual amino acids are then recorded as μg mg^{-1} of the origin sample. The addition of all individual amino acids will result in the total concentration of free amino acids in the sample.

4. FINAL REMARKS

Many alternative approaches and procedures exist to analyze amino acids by HPLC (Hodisan et al. 1998). For example, if both primary and secondary amines are to be detected, ninhydrin would be a better choice than the 4-nitrophenylisothiocyanate (NPITC) used in the protocol above to derivatize amino acids.

To minimize the considerable risk of contamination, it is critical to use high-quality chemicals and very clean, acid-washed glassware. Additionally, latex gloves should be worn at all times.

5. REFERENCES

Armstrong, S.M. & Bärlocher, F. (1989a). Adsorption and release of amino acids from epilithic biofilms in streams. *Freshwater Biology*, 22, 153-159.

Armstrong, S.M. & Bärlocher, F. (1989b). Adsorption of three amino acids to biofilms on glass-beads. *Archiv für Hydrobiologie*, 115, 391-399.

Bengtsson, G. (1988). The impact of dissolved amino acids on protein and cellulose degradation in stream waters. *Hydrobiologia*, 164, 97-102.

Cohen, S.A. (1990). Analysis of amino acids by liquid chromatography after pre-column derivatization with 4-nitrophenylisocyanate. *Journal of Chromatography*, 512, 283-290.

Craig, D., Ireland, R.J. & Bärlocher, F. (1989). Seasonal variation in the organic composition of seafoam. *Journal of Experimental Marine Biology and Ecology*, 130, 71-80.

Gessner, M.O. (1991). Differences in processing dynamics of fresh and dried leaf litter in a stream ecosystem. *Freshwater Biology*, 26, 387-398.

Hodisan, T., Culcea, M., Cimpoiu, C. & Cot, A. (1998). Separation, identification and quantitative determination of free amino acids from plant extracts. *Journal of Pharmaceutical and Biomedical Analysis*, 18, 319-323.

Rosen, H. (1957). A modified ninhydrin colorimetric analysis for amino acids. *Archives of Biochemistry and Biophysics*, 67, 10-15.

CHAPTER 11

DETERMINATION OF TOTAL
CARBOHYDRATES

SHAWN D. MANSFIELD

*Canada Research Chair in Wood and Fibre Quality, Department of Wood Science,
University of British Columbia, Vancouver, B.C., Canada, V6T 1Z4.*

1. INTRODUCTION

The plant cell wall is a composite of cellulose microfibrils intricately associated with
other polysaccharides, commonly referred to as the hemicelluloses, and an
amorphous matrix of lignin (Bidwell 1974). Many other non-structural constituents,
organic and inorganic, are present in the cell wall. Among the organic substances are
both starch and proteins, which may also be linked to carbohydrates. The
proportions of all these components varies among species, age, geographical
location, and growing conditions, and will change during decomposition due to
differential attack by various degradative enzymes (Chapter 32).

1.1. Cellulose

The architecture of the cell wall is largely determined by cellulose. The gross
physical structure and morphology of cellulose has been extensively studied, and has
been shown to be composed of long, unbranched, homopolymers of D-glucose units
linked by β-1,4-glycosidic bonds to form a linear polymer with a chain of over
10,000 glucose residues. Cellulose chains can vary in size, and associate together to
form larger marcomolecules (Hon & Shiraishi 1991).

1.2. Hemicellulose

Like cellulose most hemicelluloses function as supporting material in the plant cell
wall. These polysaccharides play an integral role in defining cell wall
characteristics, and can significantly influence the interactions between adjacent
cells. In general, hemicelluloses are low-molecular weight polysaccharides, which
are associated with the cellulose and lignin of the plant cell walls. Hemicelluloses

*M.A.S. Graça, F. Bärlocher & M.O. Gessner (eds.), Methods to Study Litter Decomposition:
A Practical Guide, 75 – 84.*

are heteropolymers constructed from a number of different residues, the most common of which are D-xylose, D-mannose, D-galactose, D-glucose, L-arabinose, D-rhamnose, D-galacturonic acid, D-glucuronic acid and 4-O-methyl-D-glucuonic acid (Fengel & Wegener 1983, Sjöström, 1993). The complexity and chemical nature of the hemicelluloses varies both between cell types and species. Hemicelluloses fall into four classes: unbranched chains, such as (1-4)-linked xylans or mannans; helical chains, such as (1-3)-linked xylans; branched chains, such as (1-4)-linked galactoglucomannans; and pectic substances, such as polyrhamnogalacturonic acid. Additionally, many hemicelluloses demonstrate substantial degrees of acetylation (Sjöström, 1993).

1.3. Starch

Like cellulose, the amylose component of starch consists of 1,4-glycosidic linked glucopyranose units, but in starch these units are α-anomers. Clearly, this stereochemical characteristic gives starch its unique physical and chemical behaviour. Amylose generally occurs as a helix in the solid state and sometimes in solution. The other major starch component in plants, amylopectin, is 1,4-α-D-glucan that is highly branched and often associated with pectic compounds (Sjöström, 1993).

1.4. Component Organization and Structure

As previously mentioned, cellulose microfibrils are usually encrusted in a lignin-hemicellulose matrix in thin sheets called lamellae, several of which make up a cell wall layer. It is believed that hemicellulose and cellulose are closely associated through hydrogen bonds, while hemicellulose and lignin are covalently linked, forming lignin-carbohydrate complexes (Eriksson et al. 1980, Watanabe et al. 1995, Newman et al. 1996, Mansfield et al. 1999). It is well recognized that the cell wall is a complex molecular entity composed primarily of polysaccharides and lignin; however, other minor structural constituents also exist, and this intricate association of macromolecules makes analyzing the individual components a difficult task.

In evaluating cell wall carbohydrates, two approaches can be taken; one involves determining total available carbohydrates (Method A) and ignores the composition and/or origin of the material, following a modified method of White & Kennedy (1986). For details of Method A, follow the orcinol total sugar method as described in Chapter 12, using the hydrolyzed sample as your unknown. The second approach (Methods B and C) is more specific and quantifies the individual carbohydrates to calculate the total amount of polysaccharide available in a substrate (Mansfield et al. 1997). However, both methods require that the lignocellulosic material first be degraded from its macromolecular structure to its individual monomeric constituents.

2. DEGRADING LIGNOCELLULOSICS: EQUIPMENT AND CHEMICALS

2.1. Equipment and Material

- Analytical balance
- Autoclave
- Desiccator (containing phosphorus pentoxide)
- Drying oven (at 105 °C)
- Grinding mill (i.e. Wiley Mill) or mortar and pestle
- Water bath (at 20 °C)
- Ice bath
- Test tubes
- Test tube rack
- Glass rods
- Septum-sealed serum bottles (150 ml)
- Medium coarseness sintered-glass crucibles (Gooch crucibles)
- Dried leaves at various stages of decay

2.2. Chemicals

- Standard sugars (e.g., glucose, galactose, arabinose, mannose, xylose)
- 72% sulphuric acid (665 ml conc. H_2SO_4 + 335 ml distilled H_2O made to 1 l in volumetric flask)
- Deionized water
- Liquid nitrogen

3. DEGRADING LIGNOCELLULOSICS: EXPERIMENTAL PROCEDURES

3.1. General

1. Grind a leaf sample to pass through a 0.5-mm mesh using a Wiley Mill. If a mill is not available, freeze dry the sample and then use a mortar and pestle to pulverize it in liquid nitrogen.
2. Dry the sample overnight at 105 °C. If samples are not used immediately, store in a desiccator containing phosphorus pentoxide.
3. Weigh out 200 mg sample into a test tube and record weight to nearest 0.1 mg.
4. Place the test tube in an ice bath.
5. Add exactly 3 ml 72% (w/w) H_2SO_4. If multiple samples are to be hydrolyzed, start time should be staggered appropriately (e.g. 10 min between samples) to ensure appropriate mixing and constant time for hydrolysis.
6. Mix and macerate the sample using a glass rod.
7. Immediately transfer the sample to a 20 °C water bath for exactly 2 h, with continuous mixing every 10 min using a glass rod.

8. After 2 h, transfer the content of the lignocellulose digestion to a serum bottle. Rinse out reaction flask with exactly 112 ml deionized H_2O and transfer washings to the serum bottle (total volume in bottle should be 115 ml; acid is now diluted to 4%) and seal with septum.

3.2. Sugar Standards

1. Prepare sugar standards indicated in Table 11.1 (sugar type and concentration will vary depending on type of lignocellulosic sample, i.e. leaf material, wood, etc.) to control for loss by decomposition during autoclaving, and to generate standard curve for quantification (this step is only required if HPLC analysis is used).

Table 11.1. Standard sugar solutions.

Standard	Composition
Sugar stock solution	(prepared in 50 ml deionized H_2O):
	Arabinose: 10 mg
	Galactose: 10 mg
	Glucose: 200 mg
	Xylose: 60 mg
	Mannose: 60 mg
High standard	30 ml sugar stock solution
	82 ml deionized H_2O
	3 ml 72% H_2SO_4
Medium standard	10 ml sugar stock solution
	102 ml deionized H_2O
	3 ml 72% H_2SO_4
Low Standard	5 ml sugar stock solution
	107 ml deionized H_2O
	3 ml 72% H_2SO_4

2. Autoclave sample(s) and sugar standards for 1 h at 121 °C.
3. Allow flasks to cool. Vacuum-filter hydrolysates through dry, pre-weighed, sintered-glass crucibles (Gooch crucibles). Ensure all solids (acid-insoluble lignin) are recovered from serum bottles by washing with small volumes (i.e. 2× 10 ml) of hydrolysate while filtering through crucible.
4. Carbohydrates in hydrolysates can now be quantified by either Method A, B or C (see below).
5. To quantify lignin (optional), wash solids in crucible with additional 200 ml warm de-ionized H_2O. Discard water and dry crucibles overnight at 105 °C. Weigh crucible and calculate lignin content by subtracting weight of pre-weighed crucible from that containing lignaceous material (this material will also contain a small amount of inorganic material, commonly referred to as ash, which can be determined by a separate analysis).

4. MONOMERIC CARBOHYDRATE DETERMINATION BY HIGH PERFORMANCE ANION EXCHANGE LIQUID CHROMATOGRAPHY (METHOD B): EQUIPMENT, AND CHEMICALS

4.1. Equipment

- HPLC (two potential method available: electrochemical detection using pulsed amperometry or refractive index detection)
- HPLC filters (0.45 μm pore size)
- HPLC vials and caps
- Disposable syringes
- Micropipettors

4.2. Chemicals

- Deionized distilled water (degassed)
- 1 M NaOH solution (degassed)
- 200 mM NaOH solution (degassed)
- Internal standard: Fucose (5 mg ml^{-1})

5. MONOMERIC CARBOHYDRATE DETERMINATION BY HIGH PERFORMANCE ANION EXCHANGE LIQUID CHROMATOGRAPHY (METHOD B): EXPERIMENTAL PROCEDURES

5.1. Sample and HPLC Preparation

1. In an Eppendorf tube add 950 μl of sample or standard and 50 μl of fucose internal standard (used to control injection volume), and mix well with a vortex. Using a disposable syringe remove sample from Eppendorf tube, and filter through 0.45 μm filter into HPLC vial, and cap.
2. Set HPLC for Electrochemical Detection using Pulsed Amperometry or Refractive Index Detection (Table 11.2).

5.2. Carbohydrate Determination

1. Run standards and samples on HPLC.
2. Normalize peak areas for internal standards (controls for injection variation).
3. Prepare standard curves by plotting area under the peak versus concentration for each sugar in the standards mixture.
4. Determine sample concentrations area using line equation ($y = mx + b$) for unknown sugars (results will be in mg ml^{-1} of each monomeric sugar).
5. Since the standard curves are generated from anhydrous sugar standards, the concentration of the unknown samples must be corrected for the conversion from polymeric to monomeric constituents (for each bond degraded during the

Table 11.2. HPLC conditions for monomeric carbohydrate determination.

Conditions	Electrochemical Detection using Pulsed Amperometry	Refractive Index Detection
Column equilibration (prior to injection)	15 min with 250 mM NaOH (degassed)	–
Mobile phase	Deionized H_2O (degassed)	Deionized H_2O (degassed)
Flow rate	1 ml min^{-1}	0.4 ml min^{-1}
Column	Dionex PA-1	Biorad HPX-87P
Detection	Electrochemical detection using pulsed amperometry (gold electrode)	Refractive index
Post column mobile phase	200 mM NaOH (at 0.5 ml min^{-1})	–
Injection volume	20 µl	20 µl

hydrolysis of polymeric carbohydrates to their corresponding monomeric moiety, include the incorporation of a water molecule). The concentration of all hexose (glucose, mannose, galactose, etc.) and pentose sugars (xylose, arabinose, etc.) is corrected by multiplying the concentration (mg ml^{-1}) by 0.9 and 0.88, respectively.

6. The total concentration in the sample is adjusted for the total volume of hydrolysate (multiple by 115 ml) and then divided by total weight of material employed (mg). The individual sugars are recorded as mg mg^{-1} of original sample. The addition of all individual monomers will result in the total carbohydrate content of each sample.

6. MONOMERIC CARBOHYDRATE DETERMINATION BY GAS CHROMATOGRAPHY (METHOD C): EQUIPMENT AND CHEMICALS

6.1. Equipment and Material

- Gas chromatograph (GC)
- Disposable filters (0.45 µm pore size)
- GC vials and caps
- Disposable syringes
- Separatory funnel
- 10 ml reaction vials with Teflon caps
- Micropipettors
- Beaker

6.2. Chemicals

- Deionized distilled water
- 3 M NH_4OH

- 2.8 M KBH_4 dissolved in 3 M NH_4OH
- Glacial acetic acid
- 1-methylimidazole
- Acetic anhydride
- Dichloromethane
- Internal standard: inositol (5 mg ml^{-1})

7. MONOMERIC CARBOHYDRATE DETERMINATION BY GAS CHROMATOGRAPHY (METHOD C): EXPERIMENTAL PROCEDURES

7.1. Sample and HPLC Preparation

1. In a 10 ml reaction vial add 200 µl of hydrolyzed sample or standard and 50 µl of inositol internal standard.
2. Add 40 µl of 3 M NH_4OH and 100 µl of the 2.8 KBH_4 solutions, and incubate for 90 min at 40 °C.
3. Stop reaction after 90 min with 100 µl of glacial acetic acid.
4. Acetylation of the sugar alcohols is completed by adding 500 µl of 1-methylimidazole and 2 ml of acetic anhydride. Vortex and allow reacting for 30 min at room temperature.
5. Add 5 ml of distilled deionized water to stop the reaction.
6. Dispense reaction mixture to a separatory funnel, where 2 ml of dichloromethane are added and mixed vigorously.
7. Allow the phases in the mixture to separate, and remove approximately 1.5 ml of the dichloromethane phase.
8. Using a disposable syringe, filter the sample is through a 0.45 µm filter into a GC vial, and cap vial.
9. Set the GC analyser to the conditions indicated in Table 11.3.

Table 11.3. Conditions for GC analysis with flame ionization detection.

Parameter	Condition
Carrier gas	Helium
Flow rate	30 cm s^{-1}
Split ratio	1:25
Column	DB-225 capillary column (30 m × 0.25 mm i.d.)
Column temperature	220 °C
Detection	Flame ionization detection (FID)
Detector temperature	240 °C
Injector temperature	240 °C
Injection volume	2 µl
Run time	30 min

7.2. Carbohydrate Determination

1. Run standards and samples on GC.
2. Normalize peak areas for internal standard (controls for injection variation).
3. Prepare standard curves by plotting area under the peak versus concentration for each sugar in the standards mixture.
4. Determine sample concentrations areas using linear equation ($y = mx+b$) for unknown sugars (results will give mg ml^{-1} of each monomeric sugar).
5. Since the standard curves are generated from anhydrous sugar standards, the concentration of the unknown samples must be corrected for the conversion from polymeric to monomeric constituents; for each bond degraded during the hydrolysis of polymeric carbohydrates to their corresponding monomeric moiety, include the incorporation of a water molecule. The concentration of all hexose (glucose, mannose, galactose, etc.) and pentose sugars (xylose, arabinose, etc.) is corrected by multiplying the concentration (mg ml^{-1}) by 0.9 and 0.88, respectively.
6. Concentration in the sample is corrected for the total volume of hydrolysate (multiple by 115 ml) and divided by total weight of material employed (mg). The individual sugars are then recorded as mg mg^{-1} of origin sample. The addition of all individual monomers will result in the total carbohydrate content of each sample.

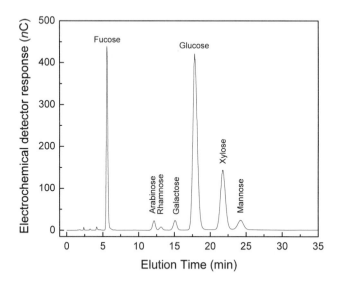

Figure 11.1. Chromatograph of neutral wood carbohydrates determined by ion exchange chromatography with electrochemical detection (pulsed amperometry).

8. REFERENCES

Bidwell, R.G.S. (1974). *Plant Physiology*. MacMillian. New York.

Eriksson, Ö., Goring, D.A.I. & Lindgren, B.O. (1980). Structural studies on the chemical bonds between lignin and carbohydrates in spruce wood. *Wood Science and Technology*, 14, 267-279.

Fengel, D. & Wegener, G. (1983). *Wood: Chemisty, Ultrastructure, Reactions*. Walter de Gruyter. New York.

Hon, D.N.-S. & Shiraishi, N. (1991). *Wood and Cellulosic Chemistry*. Marcel Dekker. New York.

Mansfield, S.D., de Jong, E. & Saddler, J.N. (1997). Cellobiose dehydrogenase, an active agent in cellulose depolymerization. *Applied and Environmental Microbiology*, 63, 3804-3809.

Mansfield, S.D., Mooney, C. & Saddler J.N. (1999). Substrate and enzyme characteristics that limit cellulose hydrolysis. *Biotechnology Progress*, 15, 804-816.

Newman, R.H., Davies, L.M. & Harris, P.J. (1996). Solid-state C13 nuclear magnetic resonance characterization of cellulose in the cell wall of *Arabidopsis thaliana* leaves. *Plant Physiology*, 111, 475-485.

Sjöström, E. (1993). *Wood Chemistry: Fundamentals and Applications*, 2nd ed. Academic Press. London.

Watanabe, T., Azuma, J. & Koshijima, T. (1995). Isolation of lignin-carbohydrate complex fragments by adsorption chromatography. *Mokuzai Gakkaishi*, 31, 52-53.

White, C.A. & Kennedy, J.F. (1986). Oligosaccharides. In: M.F. Chaplin & J.F. Kennedy (Eds.), *Carbohydrate Analysis: A Practical Approach* (pp. 37-54). IRL Press. Oxford.

CHAPTER 12

DETERMINATION OF SOLUBLE CARBOHYDRATES

SHAWN D. MANSFIELD[1] & FELIX BÄRLOCHER[2]

[1]Canada Research Chair in Wood and Fibre Quality, Department of Wood Science, University of British Columbia, Vancouver, B.C., Canada V6T 1Z4; [2]63B York Street, Department of Biology, Mount Allison University, Sackville, N.B., Canada E4L 1G7.

1. INTRODUCTION

Leaf litter and other plant detritus consist primarily of structural polysaccharides and lignin, neither of which are readily accessible to stream invertebrates (Chapters 11, 12 and 17). However, more easily digestible soluble carbohydrates, such as sucrose or glucose are also present in notable concentrations. Initially, these compounds may account for up to 16% of total dry mass (e.g. in hickory leaves; Suberkropp et al. 1976), while leaching can reduce this value by $\geq 80\%$ within a few days (Gessner 1991). The rate of leaching may be significantly influenced by treatment of the leaves before immersion in the stream (Chapter 5; Gessner 1991, Bärlocher 1997).

During decomposition, microbial enzymes attack detrital polymers, releasing a mixture of oligomeric and monomeric carbohydrates, which again are more accessible to invertebrates than the original polymers (Bärlocher & Porter 1986). In addition, fungi colonizing leaves (which can account for up to 17% of detrital dry mass at intermediate stages of decay; Gessner 1997) contain soluble carbohydrates in their mycelia.

Analysis of total soluble carbohydrates facilitates the quantification of nutritionally valuable carbon fractions of leaf material. Identification of the individual compounds, combined with analyses of hydrolyzed polysaccharides allows characterization of the course of enzymatic breakdown of these leaf constituents. The same methods can also be modified to measure activities of selected degradative enzymes present in microorganisms or invertebrates (Chapter 32).

Two approaches can be taken to analyze soluble carbohydrates. One involves determining total available carbohydrates (Method A) and ignores their composition.

M.A.S. Graça, F. Bärlocher & M.O. Gessner (eds.), Methods to Study Litter Decomposition: A Practical Guide, 85 – 90.

The procedure we present follows a modified method of White & Kennedy (1986). The second approach (Method B) is more specific. It quantifies individual monosaccharides, which can then be used to calculate the total amount of soluble carbohydrates present in a sample (Mansfield et al. 1997). Both methods require that the soluble sugars first be extracted from the lignocellulosic material; the procedure described below follows a modified protocol of Guy et al. (1984).

2. EQUIPMENT, CHEMICAL AND SOLUTIONS

2.1. Equipment and Material

- Freeze-drier
- Analytical balance
- Spectrophotometer
- Rotavap evaporator
- Desiccators (containing phosphorus pentoxide)
- Mortar and pestle
- Test tubes
- Acid-washed glass test tubes (10 ml; wash with 10% nitric acid overnight, then rinse thoroughly with distilled water)
- Test tube rack
- Hot water bath or heated test tube reactor
- Thermometer
- Ice water bath
- Vortex
- Freezer (–20 °C)
- Micropipettors
- Cuvettes (disposable ones are suitable)
- Spectrophotometer (set at 540 nm)
- Laboratory timer or stop watch
- Separatory funnel
- Aluminium foil
- High Performance Liquid Chromatograph (HPLC) with electrochemical detector using pulsed amperometry
- HPLC filters (0.45 μm pore size)
- HPLC vials and caps
- Disposable syringes

2.2. Chemicals

- Glucose
- Sucrose
- Fructose

- Liquid nitrogen
- Methanol:chloroform:water (12:5:3; v:v:v) solution
- Distilled water (degassed)
- 1 M NaOH solution (degassed)
- Internal standard: Fucose (10 mg ml^{-1})
- 10% (w:w) nitric acid
- Sugar standard: Glucose in water or appropriate buffer (e.g. 50 mM sodium phosphate or sodium acetate) with a concentration ranging from 10 to 120 µg ml^{-1}
- Freshly prepared 0.2% orcinol reagent (2 g l^{-1} orcinol dissolved in concentrated sulphuric acid); this reagent can be stored for up to one week at 4 °C

3. EXPERIMENTAL PROCEDURES

3.1. Sample Preparation

1. Wash lignocellulosic material (e.g. a leaf sample) thoroughly with distilled water.
2. Freeze-dry lignocellulosic material overnight.
3. Using a mortar and pestle grind freeze-dried material in liquid nitrogen into a powder.
4. Store ground material in sample vials in a desiccator (with phosphorus pentoxide) until extraction.
5. Weigh out 50 mg of pulverized sample into a test tube, and record weight to nearest 0.01 mg. This should be done in at least duplicate for each sample.
6. To each test tube with sample add 50 µl of fucose internal standard.
7. Add 5 ml of methanol:chloroform:water (12:5:3) solution to each test tube, mix, cover with aluminium foil, and place in a freezer for at least 12 h.
8. Remove samples from freezer and mix well with a vortex.
9. Centrifuge samples (5000 g) for 10 min at 4 °C, then pipet supernatant into separatory funnel.
10. Add 4 ml of methanol:chloroform:water (12:5:3) solution to the pellet, vortex and centrifuge for 10 min. Remove supernatant and pool with sample in separatory funnel. Repeat a second time, and pool.
11. Add 5 ml of distilled water to the separatory funnel, cap, and mix thoroughly.
12. Allow for phase separation (this could take up to 2 h).
13. Discard lower layer, and dispense aqueous top layer to evaporating flask.
14. Evaporate all solvent from flask under vacuum in a rotary evaporator at 40 °C (ensure that the water bath does not exceed 40 °C).
15. Re-suspend dried sample in 1 ml of distilled water.
16. Remove sample from evaporating flask with a disposable syringe.
17. When using the HPLC method for quantification, pass sample through 0.45 µm filter into HPLC vial and cap.

3.2. Method A – Total Soluble Carbohydrate Analysis

1. Add 0.5 ml each of the carbohydrate standard, water (blank), and/or sample to separate 10-ml test tubes with a pipettor.
2. Cover each tube with aluminium foil.
3. Immerse test tubes in ice bath (~4 °C) for 15 min.
4. Add 2 ml of orcinol reagent to the test tubes (start timer with first sample and proceed with each subsequent sample at 1 min intervals).
5. Vortex reaction mixtures vigorously and incubate in an 80 °C water bath for exactly 15 min.
6. Terminate the reaction by rapid cooling in an ice bath for 5 min.
7. Equilibrate tubes to room temperature.
8. Measure absorbance of the reaction mixture with a spectrophotometer at 540 nm (the water blank can be used for zeroing the spectrophotometer).
9. Determine total carbohydrates in the samples by reference to an appropriate standard curve generated from a standard solution (i.e. glucose).

3.3. Method B – Soluble Carbohydrate Determination by High Performance Liquid Chromatography (HPLC)

1. Set HPLC to the conditions indicated in Table 12.1.
2. Prepare sugar standards (glucose, fructose, sucrose etc.) at concentrations ranging from 0.1—2 mg ml^{-1}. A range of sugars, including sugar alcohols, can be used as standards, depending on the lignocellulosic material being analyzed or specific carbohydrates of interest.
3. Run standards and samples on HPLC to obtain chromatogram as in Fig. 12.1.
4. Normalize peaks for internal standards.
5. Prepare standard curves by plotting areas under the peaks versus concentrations for each sugar in the standards mixture.
6. Calculate concentration of each monomeric sugar in sample dry mass. The sum of all individual monomers gives the total soluble carbohydrate concentration.

Table 12.1. HPLC conditions for soluble carbohydrate determination

Parameter	Condition
Mobile phase	200 mM NaOH for 52 min; gradient from 200—420 mM NaOH from 58—80 min; 420—180 mM from 80—84 min; 180 mM from 84—100 min
Flow rate	0.4 ml min^{-1}
Column	Dionex MA-1
Column temperature	Ambient room temperature
Detection	Pulsed amperometry (using gold electrode)
Injection volume	20 µl

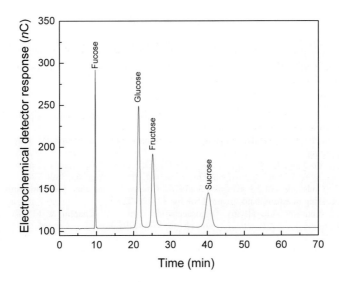

Figure 12.1. Chromatogram of soluble carbohydrate standards obtained with a Dionex HPLC system equipped with an electrochemical detector (pulsed amperometry) and the conditions described for Method B.

4. FINAL REMARKS

Samples analyzed by either method may require dilution if the unknown samples give absorbance or detection readings greater than the highest value obtained while generating the standard curve. Should this occur, dilute samples, record dilution volume, and repeat the analysis.

As a spectrophotometric assay, Method A is subject to interference from particles or air bubbles in sample or reaction solutions. Different sugars give a different quantitative response (e.g. glucose $\neq$ xylose). Therefore, choice of the standards will depend on the carbohydrate composition of the sample. Sensitivity of Method A is approximately 5—10 μg ml^{-1} carbohydrate (1 ml sample required); that of Method B $\geq$1 μg ml^{-1} carbohydrate.

Orcinol is a harmful substance and special care needs to be taken when handling it, especially when it is made up as a solution in concentrated sulphuric acid. As such, laboratory coats, protective eyewear and gloves are required.

4. REFERENCES

Bärlocher, F. (1997). Pitfalls of traditional techniques when studying decomposition in aquac habitats. *Limnetica*, 13, 1-11.

Bärlocher, F. & Porter, C.W. (1986). Digestive enzymes and feeding strategies of three stream invertebrates. *Journal of the North American Benthological Society*, 5, 58-66.

Gessner, M.O. (1991). Differences in processing dynamics of fresh and air-dried leaf litter in a stream ecosystem. *Freshwater Biology,* 26, 387-398.

Gessner, M.O. (1997). Fungal biomass, production and sporulation associated with particulate organic matter in streams. *Limnetica*, 13, 33-44.

Guy, R.D., Warne, P. G. & Reid, D.M. (1984). Glycinebetaine content of halophytes: Improved analysis by liquid chromatography and interpretation of results. *Physiologia Plantarum*, 61, 195-202.

Mansfield, S.D., de Jong, E. & Saddler, J.N. (1997). Cellobiose dehydrogenase, an active agent in cellulose depolymerization. *Applied and Environmental Microbiology*, 63, 3804-3809.

Suberkropp, K., Godshalk, G.L. & Klug, M.J. (1976). Changes in the chemical composition of leaves during processing in a woodland stream. *Ecology*, 57, 720-727.

White, C.A. & Kennedy, J.F. (1986). Oligosaccharides. In: M.F. Chaplin & J.F. Kennedy (eds.) *Carbohydrate Analysis: A Practical Approach* (pp. 37-54) IRL Press Ltd. Oxford.

CHAPTER 13

TOTAL LIPIDS

MARK O. GESSNER[1] & PAUL T.M. NEUMANN[2]

[1]*Department of Limnology, EAWAG, Limnological Research Centre, 6047 Kastanienbaum, Switzerland; [2]Limnologische Fluss-Station des Max-Planck-Instituts für Limnologie, 36110 Schlitz, Germany; present address: Friedrich-Kreusch-Weg 41, 40764 Langenfeld, Germany.*

1. INTRODUCTION

Lipids are a major class of chemical compounds in plant tissues that have rarely been considered in litter decomposition studies and assessments of litter nutritional quality for aquatic detritivores. This neglect partly reflects the fact that emphasis in the chemical characterization of decomposing litter has been placed on nutrients, particularly in terms of nitrogen (Enriquez et al. 1993), and refractory litter constituents such as lignin (e.g. Gessner & Chauvet 1994, Palm & Rowland 1997). The currently available data show, however, that lipids can be a sizeable fraction of plant litter (Table 13.1), suggesting that information on lipid content may be useful when modelling decomposition as a function of chemical litter composition (e.g. Moorhead et al. 1999).

There is evidence, moreover, that lipids can provide critical cues to detritivore feeding (Anderson & Cargill 1987). For example, the sequence of food preference

Table 13.1. Total average lipid content of undecomposed leaf litter of various tree species. AFDM = ash-free dry mass.

Leaf species	Common name	Lipids (% AFDM)	Reference
Acacia melanoxylon	Blackwood acacia	1.9[a]	1
Carya glabra	Hickory	5.2	2
Eucalyptus viminalis	Manna eucalyptus	12[a]	1
Liquidamber styraciflua	Sweetgum	1.7[a,b]	3
Pomaderris aspera	Hazel pomaderris	3.8[a]	1
Quercus alba	White oak	4.9	2
Quercus nigra	Water oak	1.0[a,b]	3

[a]AFDM assumed to be 90% of dry mass. [b]Based on fatty acid content of saponified methanol extracts. 1 = Campbell et al. (1992); 2 = Suberkropp & Klug (1976); 3 = Mills et al. (2001)

M.A.S. Graça, F. Bärlocher & M.O. Gessner (eds.), Methods to Study Litter Decomposition: A Practical Guide, 91 – 96.

of two detritivores (*Gammarus tigrinus* and *Pycnopsyche guttifer*) for certain combinations of fungal species grown on leaves could be reproduced by applying lipid extracts of the fungi to uncolonized leaves (Rong et al. 1995). For holometabolic insects with limited food acquisition during the adult stage, lipid content may be a particularly important litter quality attribute, because late instars of these invertebrates benefit from consuming food with a high energy content to build up energy reserves before emergence. Support for this hypothesis has come from a detritivorous caddisfly, the limnephilid *Clistoronia magnifica* (Hanson et al. 1983, Cargill et al. 1985a, b).

Bulk lipid analyses have classically adopted a gravimetric approach (Suberkropp et al. 1976). Lipids are extracted from the tissue with an apolar solvent, which is then evaporated and the dried residue is weighed. The main limitation of this approach is that relatively large sample sizes are needed for accurate analyses. An attractive alternative is a spectrophotometric assay that allows analysis of small samples. Zöllner & Kirsch (1962) described such a method for analyzing blood lipids. This method, known as the sulphophosphovanillin assay, has later been applied to estimating lipid contents of algae (Rausch 1981, Ahlgren & Merino 1991), aquatic invertebrates (Barnes & Blackstock 1973, Meyer & Walther 1989), and fine-particulate organic matter (Neumann 1995). It is based on the reaction of lipid degradation products with aromatic aldehydes, which results in a red coloration that can be quantified at 528 nm (Zöllner & Kirsch 1962). With particulate organic matter, it is essential to extract the lipids from the bulk sample before performing the sulphophosphovanillin assay. This is because strong interference by nonlipid compounds (e.g. carbohydrates) results in high nonspecific absorbance after heating the sample in sulphuric acid and addition of the vanillin reagent, and thus in unreliable results (Ahlgren & Merino 1991, Neumann 1995).

The protocol described here has been adopted from Neumann (1995). Although developed for fine-particulate organic matter, it has also proved reliable for decomposing leaves from streams (M.O. Gessner, unpubl. data).

2. EQUIPMENT, CHEMICALS AND SOLUTIONS

2.1. Equipment and Material

- Freeze-dryer
- Mill
- Analytical balance
- Centrifuge tubes (12 ml, preferably pressure resistant, with Teflon-lined screw-caps)
- Centrifuge tubes (4 ml)
- Pipettes allowing precise pipetting of solvents with low viscosity (e.g. Eppendorf Multipette® or Varipette®)
- Standard laboratory centrifuge
- Evaporating centrifuge (e.g. SpeedVac concentrator SPD131DDA, Thermo Electron Corp., Woburn, MA, USA)

- Vials (4 ml, with Teflon-lined screw caps)
- Vortex mixer
- Pear-shaped glass bulbs or marbles (only needed if centrifuge tubes do not resist high pressure)
- Water or dry baths (20 and 100 °C)
- Timer
- Spectrophotometer (set at 528 nm)
- Volumetric flasks

2.2. Chemicals

- Chloroform ($CHCl_3$), for residual analysis
- Methanol (CH_3OH), for residual analysis
- Deionized water (e.g. Nanopure®)
- NaCl, reagent grade
- Concentrated sulphuric acid (H_2SO_4, 95—97%), reagent grade
- Concentrated phosphoric acid (H_3PO_4, 85%), reagent grade
- Vanillin
- Cholesterol (5-cholesten-3β-ol)

2.3. Solutions

- 0.9% (w:v) NaCl solution
- Phosphoric acid-vanillin reagent: 20 ml 0.6% (w:v) vanillin solution and 85% H_3PO_4 in a total volume of 100 ml
- Standard solutions: Cholesterol standards in chloroform at concentrations ranging from 10 to 100 mg ml^{-1}.

3. EXPERIMENTAL PROCEDURES

3.1. Lipid Extraction

1. Freeze-dry leaves and grind to powder that passes trough a 0.5-mm mesh screen.
2. Weigh out 25 mg of sample to the nearest 0.1 mg in a 12-ml screw-cap centrifuge tube with a Teflon-lined cap.
3. Add 7 ml chloroform:methanol (2:1, v:v).
4. Shake for 1 min, then let stand for 2 h, with shaking for 1 min every 30 min.
5. Centrifuge for 1 min at about 3000 *g*, rinse tube walls to suspend any adhering particles, then centrifuge for another 10 min to separate particles from the lipid extract.
6. Transfer 5 ml of the lipid extract to a clean tube containing 1 ml of 0.9% NaCl solution.
7. Shake for 1 min, then centrifuge for 10 min at about 3000 *g* to separate phases.

8. Remove and discard upper aqueous phase.
9. Rinse inner walls of the tube twice with 1 ml chloroform:methanol:water (3:48:47, v:v:v).
10. Remove rinsing solution.
11. Evaporate sample to dryness in a SpeedVac concentrator at about 45 °C.
12. Transfer residue with 2 ml chloroform to a clean 4-ml screw-cap vial with Teflon-lined cap.
13. Evaporate sample again to dryness in the SpeedVac concentrator.
14. Redissolve residue in 1 ml chloroform, close vial tightly and run spectrophotometric assay or store at –20 °C until analysis.
15. Run control without sample material in the same way.

3.2. Spectrophotometric Analysis

1. Place 100 µl of the lipid extract in a 12-ml test tube.
2. Evaporate solvent in SpeedVac concentrator at about 45 °C.
3. Add 200 µl of conc. H_2SO_4 and vortex.
4. Close tube tightly or, if not pressure-resistant, cover it with a pear-shaped glass bulb (or marble), and heat for 10 min to 100 °C in a water or dry bath.
5. Let cool for 5 min in a water bath at 20 °C.
6. Add 2.5 ml of H_3PO_4-vanillin reagent and vortex.
7. Measure absorbance after 60—65 min at 528 nm.
8. Run cholesterol standards in the same way.
9. Calculate lipid content as cholesterol equivalents from absorbance reading of sample and standard curve.

4. FINAL REMARKS

The time course of colour development depends strongly on the ratio of sulphuric acid to the H_3PO_4-vanillin solution (Neumann 1995). It is essential, therefore, to keep this ratio strictly constant in a given sample series (e.g., at 1:12.5 as in the protocol described above).

At a 1:12.5 ratio of sulphuric acid to the H_3PO_4-vanillin reagent, absorbance readings must be taken 60—65 min after addition of reagent to the sample. Earlier or later readings result in an underestimation of lipid contents.

Since lipids are a highly heterogeneous class of molecules, choice of an appropriate standard is critical to facilitate accurate quantitative estimates. Neumann (1995) found that cholesterol and a lipid extract from FPOM collected in a stream gave identical responses with the protocol described here. However, use of a specific standard (i.e., a lipid solution from a representative sample, with the lipid content determined gravimetrically) may be preferable when precise information about the absolute magnitude of lipid concentrations is required. For finely ground FPOM, the efficiency of lipid extraction in a single step as described above was 93% compared to three successive extraction steps (Neumann 1995).

4. REFERENCES

Ahlgren, G. & Merino, L. (1991). Lipid analysis of freshwater microalgae: A method study. *Archiv für Hydrobiologie*, 121, 295-306.

Anderson, N.H. & Cargill, A.S. (1987). Nutritional ecology of aquatic detritivorous insects. In: F. Slansky Jr. & J.G. Rodriguez (eds.), *Nutritional Ecology of Insects, Mites, Spiders, and Related Invertebrates* (pp. 903-925). Jon Wiley & Sons. New York.

Barnes, H. & Blackstock, J. (1973). Estimates of lipids in marine animals and tissues: Detailed investigation of the sulphophosphovanillin method for total lipids. *Journal of Experimental Marine Biology and Ecology*, 12, 103-118.

Campbell, I.C., James, K.R., Hart, B.T. & Devereaux, A. (1992). Allochthonous coarse particulate organic material in forest and pasture reaches of two south-eastern Australian streams. 2. Litter processing. *Freshwater Biology*, 27, 353-365.

Cargill, A.S. II, Cummins, K.W., Hanson, B.J. & Lowry, R.R. (1985a). The role of lipids as feeding stimulants for shredding aquatic insects. *Freshwater Biology*, 15, 455-464.

Cargill, A.S. II, Cummins, K.W., Hanson, B.J. & Lowry, R.R. (1985b). The role of lipids, fungi, and temperature in the nutrition of a shredder caddisfly, *Clistoronia magnifica*. *Freshwater Invertebrate Biology*, 4, 64-78

Enríquez, S., Duarte, C.M. & Sand-Jensen, K. (1993). Patterns in decomposition rates among photosynthetic organisms: the importance of detritus C:N:P content. *Oecologia*, 94, 457-471.

Gessner, M.O. & Chauvet, E. (1994). Importance of stream microfungi in controlling breakdown rates of leaf litter. *Ecology*, 75, 1807-1817.

Hanson, B.J., Cummins, K.W., Cargill, A.S. II & Lowry, R.R. (1983). Dietary effects on lipid and fatty acid composition of *Clistoronia magnifica* (Trichoptera, Limnephilidae). *Freshwater Invertebrate Biology*, 2, 2-15.

Meyer, E. & Walther, A. (1989). Methods for the estimation of protein, lipid, carbohydrate and chitin levels in fresh water invertebrates. *Archiv für Hydrobiologie*, 113, 161.177.

Mills, G.L., McArthur, J.V., Wolfe, C., Aho, J.M. & Rader, R.B. (2001). Changes in fatty acid and hydrocarbon composition of leaves during decomposition in a southeastern blackwater stream. *Archiv für Hydrobiologie* 152: 315-328.

Moorhead, D.L., Currie, W.S., Rastetter, E.P., Parton, W.J. & Harmon, M.E. (1999). Climate and litter quality controls of decomposition: an analysis of modelling approaches. *Global Biogeochemical Cycles*, 13, 575-589.

Neumann, P. (1995). Untersuchungen zur Nahrungsqualität von benthischem feinpartikulärem Detritus für Feinpartikelsammler, unter dem Aspekt seiner biochemischen Zusammensetzung im Breitenbach. Unpublished PhD Thesis, University of Marburg, Germany, 135 pp.

Palm, C.A. & Rowland, A.P. (1997). A minimum data set for characterization of plant quality for decomposition. In: G. Cadish & K.E. Giller (eds.), *Driven by Nature: Plant Litter Quality and Decomposition* (pp. 379-391). CAB International. Wallingford.

Rong, Q., Sridhar, K.R. & Bärlocher, F. (1995). Food selection in three leaf-shredding stream invertebrates. *Hydrobiologia*, 316, 173-181.

Suberkropp, K., Godshalk, G.L. & Klug, M.J. (1976). Changes in the chemical composition of leaves during processing in a woodland stream. *Ecology*, 57, 720-727.

Zöllner, N. & Kirsch, K. (1962). Über die quantitative Bestimmung von Lipoiden (Mikromethode) mittels der vielen natürlichen Lipoiden (allen bekannten Plasmalipoiden) gemeinsamen Sulphophosphovanillin-Reaktion. *Zeitschrift für die gesamte experimentelle Medizin*, 135, 545-561.

CHAPTER 14

TOTAL PHENOLICS

FELIX BÄRLOCHER[1] & MANUEL A.S. GRAÇA[2]

[1]63B York Street, Department of Biology, Mt. Allison University, Sackville, N.B., Canada, E4L 1G7; [2]Departmento de Zoologia, Universidade de Coimbra, 3004-517, Coimbra, Portugal.

1. INTRODUCTION

Phenolics are a heterogeneous group of natural substances characterized by an aromatic ring with one or more hydroxyl groups. The number of these compounds identified to date exceeds 100,000 (Waterman & Mole 1994). Phenolics may occur as monomers with one hydroxyl group (e.g. ferulic acid). Compounds with several phenolic hydroxyl substituents are referred to as polyphenolics. Among these, tannins (subdivided into phlorotannins, hydrolyzable and condensed tannins) are of particular interest because of various demonstrated or posited ecological effects (Zucker 1988; Chapter 15). In particular, phenolics play a major role in the defence against herbivores and pathogens (Waterman & Mole 1994, Lill & Marquis 2001). In addition, some phenolics such as anthocyanins may prevent leaf damage resulting from exposure to excessive light (Gould & Lee 2002). Since the bulk of phenolics remains present during leaf senescence and after death, these compounds may also affect microbial decomposers (Harrison 1971) and therefore delay microbial decomposition of plant litter (Zucker 1988, Salusso 2000). The amount of phenolics in plant tissues varies with leaf species, age and degree of decomposition. Values for selected plants are summarized in Table 14.1.

A first step in many studies assessing the ecological effects of phenolics is an estimate of the total concentration of phenolic hydroxyl groups. The most commonly used assay for that purpose was originally designed to quantify the phenolic amino acid tyrosine (Folin & Denis 1912). Folin & Ciocalteu (1927) made the assay more sensitive and less prone to formation of precipitates. Preparation of the Folin-Denis or Folin-Ciocalteu reagent is relatively time-consuming, but these reagents are now commercially available (Waterman & Mole 1994). Here we present the procedure introduced by Folin & Ciocalteu (1927).

M.A.S. Graça, F. Bärlocher & M.O. Gessner (eds.), Methods to Study Litter Decomposition: A Practical Guide, 97 – 100.

Table 14.1. Phenolics concentrations in terms of tannic acid or ferulic acid equivalents for selected plant tissues, including senescent leaves (s), live (l) and yellow-green to brown-dead grass leaves (y)

Species	Common name	Phenolics (% leaf dry mass)	Reference
Spartina alterniflora (y)	Smooth cordgrass	0.4—1.5	1
		0.2—1.2	2
Alnus glutinosa (s)	Alder	2.7	3
		6.6	4
		6.8—7.6	5
Sapium sebiferum (l, s)	Chinese tallow tree	3.0	6
Eucalyptus globulus (s)	Eucalyptus	6.4	3
		9.8	4
Fagus sylvatica (s)	Beech	8.0	7
Carya glabra (s)	Hickory	9.1	8
Quercus alba (s)	Oak	16.2	8
Acer saccharum (s)	Sugar maple	15	7

1 = Graça et al. (2000); 2 = Bärlocher & Newell (1994); 3 = Pereira et al. (1998); 4 = Bärlocher et al. (1995); 5 = Gessner (1991); 6 = Cameron & LaPoint (1978); 7 = Graça & Bärlocher (1998); 8 = Suberkropp & Klug (1976).

2. EQUIPMENT, CHEMICALS AND SOLUTIONS

2.1. Equipment and Material

- Eppendorf pipettes
- Vortex
- Refrigerator
- Dried leaves
- Mill or mortar and pestle
- Analytical balance (±0.1 mg precision)
- Eppendorf tubes
- Centrifuge
- Spectrophotometer

2.2. Chemicals

1. Tannic acid standard
2. Acetone
3. Deionized water
4. 2% Na_2CO_3
5. 0.1 N NaOH
6. Folin-Ciocalteu reagent (e.g. Sigma F-9252; diluted 1:2 with deionized water)

3. EXPERIMENTAL PROCEDURES

3.1. Calibration

1. Prepare a stock solution of 25 mg tannic acid in 100 ml of acetone (30% water, 70% acetone).
2. Transfer 0, 0.2, 0.4, 0.6, 0.8, 1.0 ml of the stock solution into 6 Eppendorf tubes and add 1.0, 0.8, 0.6, 0.4, 0.2 and 0 ml of distilled water, respectively. Mix with vortex.
3. Add 5 ml of 2% Na_2CO_3 in 0.1 N NaOH and mix.
4. After 5 min, add 0.5 ml of Folin-Ciocalteu reagent and mix.
5. After 120 min, read absorbance at 760 nm.
6. Plot tannic acid concentration vs. absorbance. The relationship should be linear.

3.2. Measurement

1. Grind up dried leaves. Use powder passing trough a 0.5-mm mesh size screen.
2. Weigh out approximately 100 mg portions of the ground leaves and transfer to Eppendorf tubes.
3. Extract phenolics in 5 ml of 70% acetone for 1 h at 4 °C.
4. Centrifuge (10,000—20,000 g, 10—20 min).
5. Take 0.5 ml of the supernatant (or another value between 0.1 and 0.8), make up to 1 ml with distilled water as above.
6. Add Na_2CO_3 and Folin-Ciocalteu reagent as above.
7. After 120 min, read absorbance at 760 nm.
8. Based on the standard curve, determine tannic acid equivalents per mg of leaf powder. Remember that in Step 5, only a fraction (0.5 ml) of the sample was used.

4. REFERENCES

Bärlocher, F., Canhoto, C. & Graça, M.A.S. (1995). Fungal colonization of alder and eucalypt leaves in two streams in Central Portugal. *Archiv für Hydrobiologie*, 133, 457-470.

Bärlocher, F. & Newell S.Y. (1994). Phenolics and protein affecting palatability of *Spartina* leaves to the gastropod *Littoraria irrorata*. *P.S.Z.N.I.: Marine Ecology*, 15, 65-75.

Cameron, G.N. & LaPoint, T.W. (1978). Effects of tannins on the decomposition of Chinese tallow leaves by terrestrial and aquatic invertebrates. *Oecologia*, 32, 349-366.

Folin, O. & Ciocalteu, V. (1927). On tyrosine and tryptophane determination in proteins. *Journal of Biological Chemistry*, 27, 239-343.

Folin, O. & Denis, W. (1912). Tyrosine in proteins as determined by a new colorimetric method. *Journal of Biological Chemistry*, 12, 245-251.

Gould, K.S. & Lee, D.W. (eds.) (2002). *Anthocyanins and Leaves. The Function of Anthocyanins in Vegetative Organs. Advances in Ecological Research*, Vol. 37. Academic Press. London.

Gessner M.O. 1991. Differences in processing dynamics of fresh and dried leaf litter in a stream ecosystem. *Freshwater Biology*, 26, 387-398.

Graça, M.A.S. & Bärlocher, F. (1998). Proteolytic gut enzymes in *Tipula caloptera* – Interaction with phenolics. *Aquatic Insects*, 21, 11-18

Graça, M.A.S., Newell, S.Y. & Kneib, R.T. (2000). Grazing rates of organic living fungal biomass of decaying *Spartina alterniflora* by three species of salt-marsh invertebrates. *Marine Biology*, 136, 281-289.

Harrison, A.F. (1971). The inhibitory effect of oak leaf litter tannins on the growth of fungi, in relation to litter decomposition. *Soil Biology and Biochemistry*, 3, 167-172.

Lill, J.T. & Marquis, R.J. (2001). The effects of leaf quality on herbivore performance and attack from natural enemies. *Oecologia*, 126, 418-428.

Pereira, A.P., Graça, M.A.S. & Molles, M. (1998). Leaf decomposition in relation to litter physico-chemical properties, fungal biomass, arthopod colonization, and geographical origin of plant species. *Pedobiologia*, 42, 316-327.

Rosset, J., Bärlocher, F. & Oertli, J.J. (1982). Decomposition of conifer needles and leaf litter in two Black Forest and two Swiss Jura streams. *Internationale Revue der gesamten Hydrobiologie*, 67, 695-711.

Salusso, M.M. (2000). Biodegradation of subtropical forest woods from north-west Argentina by *Pleurotus laciniatocrenatus*. *New Zealand Journal of Botany*, 38, 721-724.

Suberkropp, K., Godshalk G.L. & Klug, M.J. (1976) Changes in the chemical composition of leaves during processing in a woodland stream. *Ecology*, 57, 720-727.

Waterman, P.G. & Mole, S. (1994). *Analysis of Phenolic Plant Metabolites*. Blackwell Scientific Publications. Methods in Ecology. Oxford.

Zucker, W.V. (1983). Tannins: does structure determine function? An ecological perspective. *American Naturalist*, 121, 335-365.

CHAPTER 15

RADIAL DIFFUSION ASSAY FOR TANNINS

MANUEL A.S. GRAÇA[1] & FELIX BÄRLOCHER[2]

[1]*Departamento de Zoologia, Universidade de Coimbra, 3004-517, Coimbra, Portugal;* [2]*63B York Street, Depatment of Biology, Mt. Allison University, Sackville, N.B., Canada, E4L 1G7.*

1. INTRODUCTION

Few classes of secondary metabolites have been studied as intensively as tannins (Waterman & Mole 1994). They are often referred to as polyphenolics, although not all polyphenolics are tannins. There are three chemically distinct types of tannins: phlorotannins (restricted to brown algae), hydrolysable tannins (some green algae and angiosperms) and condensed tannins (most widely distributed group; Waterman & Mole 1994; Chapter 14). Swain (1979) defines tannins as polymeric compounds (i) having molecular weights between 1000 and 3000; (ii) having sufficient phenolic hydroxyl groups to complex with proteins and other macromolecules possessing carbonyl and amino groups; (iii) forming hydrogen bonds with macromolecules that are susceptible to auto-oxidation to form covalent linkages.

Total phenolics or tannins are often negatively correlated with feeding preferences of vertebrates and invertebrates and with microbial decomposition (Rosset et al. 1982, Pennings et al. 2000). They also reduce nutrient extraction of ingested food, or, in the long term, interfere with reproduction (Harrison 1971, Neuhauser & Hartenstein 1978, Zimmer & Topp 2000, Simpson & Raubenheimer 2001). By retarding microbial degradation and invertebrate feeding, these compounds may influence decomposition rates and therefore rates at which nutrients are recycled.

There are two main approaches to measure tannins based on their interaction with proteins (Waterman & Mole 1994). One takes advantage of their ability to inhibit enzyme catalyzed reactions. The other assesses the capability of tannins to bind to and precipitate proteins. This capability is exploited in the radial diffusion assay (RDA) introduced by Hagerman (1987) and described here. A standard protein is dissolved in an agar gel. A well is then punched in the gel and a known amount of plant extract added to it. The tannins in the plant extract will bind to and thus precipitate the protein. The phenol-protein complex appears as ring with an area proportional to the 'tanning' or 'protein-precipitating' activity (Fig. 15.1)

M.A.S. Graça, F. Bärlocher & M.O. Gessner (eds.), Methods to Study Litter Decomposition: A Practical Guide, 101 – 108.

The radial diffusion assay does not quantify the total amount of tannins, but their ability to bind proteins, and therefore addresses only one aspect of the biological significance of these compounds. Some representative results are shown in Table 15.1.

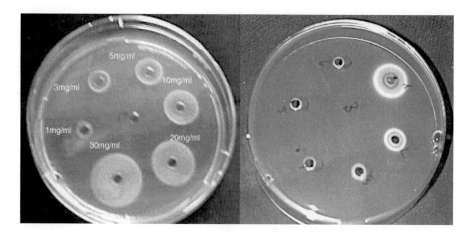

Figure 15.1. Plaques on an agarose gel resulting from the application of different amounts of commercial tannic acid (left) and different amounts of tannin in a leaf extract (right) Photo D. Steiner.

Table 15.1. Tannic acid equivalents of leaves (% of dry mass) determined with the radial diffusion assay by Hagerman (1987).

Species	Common name	Tannic acid equivalents	Reference
Alnus glutinosa	Alder	1.5	1
		6.5—7.6	2
		5.2	3
Corylus avellana	Hazel	3.6	3
Eucalyptus globulus	Eucalyptus	3.5	1
Fagus sylvatica	Beech	2.9	3
Fraxinus excelsior	Ash	0.0	3
Platanus hybrida	London plane	2.7	3
Prunus avium	Cherry	3.1	3
Quercus ilex	Evergreen oak	6.7	3

1 = Bärlocher et al. (1995); 2 = Gessner (1991); 3 = Gessner & Chauvet (1994).

2. EQUIPMENT, CHEMICALS AND SOLUTIONS

2.1. Equipment and Materials

- Hot plate, with magnetic stir bar
- Eppendorf pipettes
- Petri plates, 8.5 cm diameter
- Cork borer (2—4 mm diameter)
- Refrigerator
- Drying oven (20 °C)
- Mill or mortar and pestle
- Analytical balance (±0.1 mg precision)
- Eppendorf tubes
- Centrifuge (10,000—20,000 g)
- Water bath

2.2. Chemicals

- Tannic acid standard
- Bovine Serum Albumin (BSA)
- Agarose
- Ascorbic acid
- Acetone
- Solution A: 0.2 M acetic acid (11.55 ml glacial acid in 1000 ml deionized water)
- Solution B: 0.2 M sodium acetate (16.4 g sodium acetate in 1000 ml deionized water)
- Solution C: Combine 74 ml of solution A and 176 ml of solution B and adjust to 500 ml with deionized water.

3. EXPERIMENTAL PROCEDURES

3.1. Preparation of Agarose Plates

1. Dissolve 5 g of agarose and 5.3 mg ascorbic acid (= 60 µM) in 500 ml of Solution C on a hotplate. Stir continuously.
2. Cool down to 45 °C in a water bath.
3. Add BSA to a final concentration of 0.1 % while stirring (500 mg in 500 ml).
4. Dispense in 9.5 ml portions into Petri plates while avoiding foam formation and bubbles in the gel.
5. After gelling, punch out 2—4 mm wells.
6. Store plates at 4 °C.

3.2. Calibration

1. Dissolve 250 mg tannic acid in 25 ml of acetone (30 % water, 70 % acetone).
2. Add 9, 18, 27, 36, 45, or 54 µl of this solution to different wells. This is best done by repeatedly adding portions of 9 µl, allowing 10 min between each application.
3. After incubation for 3—4 days at 20 °C, determine the area of precipitation (Fig. 15.1). Measure the diameter of the ring twice at right angles and calculate the surface area from the average diameter. A plot of ring area vs. amount of tannic acid in well (subtract area of well) should give a linear relationship.

3.3. Measurement

1. Grind up dried leaves. For the assay use powder passing trough a 0.5 mm mesh size screen.
2. Weigh approximately 100 mg portions of ground leaves and transfer these samples to Eppendorf tubes.
3. Extract the tannin content by adding 0.5 ml (or 1 ml) of 70% acetone to the tubes. Leave for 1 h in the refrigerator for extraction.
4. Centrifuge (10,000—20,000 g, 10—20 min).
5. Transfer aliquots of the supernatant (generally 4×9 µl) to the wells in the Petri plates.
6. Incubate for 3—4 days at 20 °C; then measure the area of precipitation as for the standards above.
7. Express the results in tannic acid equivalents by comparing them to the protein-precipitating capacity of tannic acid.

4. REFERENCES

Bärlocher, F., Canhoto, C. & Graça, M.A.S. (1995). Fungal colonization of alder and eucalypt leaves in two streams in Central Portugal. *Archiv für Hydrobiologie*, 133, 457-470.

Gessner, M.O. (1991). Differences in processing dynamics of fresh and air-dried leaf litter in a stream ecosystem. *Freshwater Biology*, 26, 387-398.

Gessner, M.O. & Chauvet, E. (1994). Importance of stream microfungi in controlling breakdown rates of leaf litter. *Ecology*, 78, 1807-1817.

Hagerman, A.E. (1987). Radial diffusion method for determining tannin in plant extracts. *Journal of Chemical Ecology*, 13, 437-449.

Harrison, A.F. (1971). The inhibitory effect of oak leaf litter tannins on the growth of fungi, in relation to litter decomposition. *Soil Biology and Biochemistry*, 3, 167-172.

Neuhauser, E.F. & Hartenstein, R. (1978). Phenolic content and palatability of leaves and wood to soil isopods and diplopods. *Pedobiologia*, 18, 99-109.

Pennings, S.C., Carefoot, T.H., Zimmer, M., Danko, J.P. & Ziegler, A. (2000). Feeding preferences of supralittoral isopods and amphipods. *Canadian Journal of Zoology*, 78, 1918-1929.

Rosset, J., Bärlocher, F. & Oertli, J.J. (1982). Decomposition of conifer needles and leaf litter in two Black Forest and two Swiss Jura streams. *Internationale Revue der gesamten Hydrobiologie*, 67, 695-711.

Simpson, S.J. & Raubenheimer, D. (2001). The geometric analysis of nutrient-allelochemical interactions: A case study using locusts. *Ecology*, 82, 422-439.

Swain, T. (1979). Tannins and lignins. In: G. Rosenthal & D.H. Hanzen (eds.). *Herbivores: their interaction with secondary plant metabolites*. pp. 657-682. Academic Press. New York.

Waterman, P.G. & Mole, S. (1994). *Analysis of Phenolic Plant Metabolites*. Blackwell Scientific Publications, Methods in Ecology. Oxford.

Zimmer, M. & Topp, W. (2000). Species-specific utilization of food sources by sympatric woodlice (Isopoda: Oniscidea). *Journal of Animal Ecology*, 69, 1071-1082.

CHAPTER 16

ACID BUTANOL ASSAY FOR PROANTHOCYANIDINS (CONDENSED TANNINS)

MARK O. GESSNER & DANIEL STEINER

Department of Limnology, EAWAG, Limnological Research Centre, 6047 Kastanienbaum, Switzerland.

1. INTRODUCTION

Tannins are a major class of secondary metabolites that are widespread in plants (Waterman & Mole 1994, Kraus et al. 2003a). They are water-soluble polyphenolics with molecular weights typically ranging from 1000 to 3000 (Swain 1979). By definition, tannins are capable of complexing and subsequently precipitating proteins (cf. Chapter 15), and they can also bind to other macromolecules (Zucker 1983). Two main, chemically distinct groups are commonly distinguished in vascular plants: hydrolysable tannins, which are further divided into the gallotannins and ellagitannins, and condensed tannins, or proanthocyanidins, which cannot be hydrolyzed (Waterman & Mole 1994, Hättenschwiler & Vitousek 2000). Proanthocyanidins are the most widely distributed tannins in woody plants. They are usually also the most abundant group. Their diversity both within and among species is remarkable; however, the polymeric structures of proanthocyanidins can be derived from relatively few building blocks of low-molecular weight compounds. The most important monomers are flavan-3-ols such as catechin, epicatechin, gallocatechin and epigallocatechin; they react with one another in various ways, leading to either linear or branched polymers (Fig. 16.1).

Discussions on the ecological functions of tannins have mainly revolved around their capacity to bind to proteins and precipitate them (Zucker 1983). Both vertebrate and invertebrate herbivores can be affected. Herbivores also tend to prefer diets with low tannin concentrations, suggesting that tannins act as feeding deterrents to these consumers, although evidence supporting this tenet is inconclusive (Ayres et al. 1997). A range of additional general ecological functions at both the organismic and ecosystem level have been proposed (Hättenschwiler & Vitousek 2000, Kraus et al. 2003a). These include the role of tannins as antioxidants, mediators of nutrient availability in soils, and regulating factors of

107

M.A.S. Graça, F. Bärlocher & M.O. Gessner (eds.), Methods to Study Litter Decomposition: A Practical Guide, 107 – 114.

litter decomposition. In addition, as Zucker (1983) pointed out more than two decades ago, the chemical structure of tannins suggests that there is tremendous scope for specific chemical interactions of tannins both within organisms and in ecosystems. This view of multiple ecological roles for tannins is now widely accepted but data that would allow assembling a clear overall picture of tannin function are still limited (Hättenschwiler & Vitousek 2000, Kraus et al. 2003a).

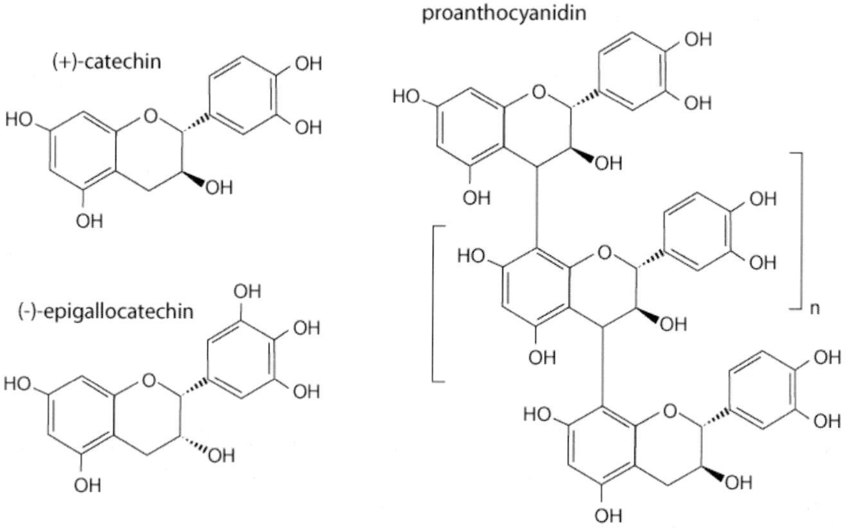

Figure 16.1. Flavan-3-ols (+)-catechin and (-)-epigallocatechin, examples of monomeric precursors that polymerize to form macromolecular products such as linear proanthocyanidins composed of monomeric flavanoid units connected by C4-C8 linkages.

If tannins remain in leaves following abscission (Table 16.1), similar mechanisms as in plant-herbivore interactions would be expected for trophic interactions between leaf litter and detritivores (e.g. Stout 1989, Ostrofsky 1997, Kraus et al. 2003a), with consequent effects on detritivore performance (Zimmer et al. 2002). There is evidence, moreover, that tannins interact with microbial decomposers (Kraus et al. 2003a), indicating that there is significant potential for tannins to affect litter decomposition in both terrestrial (Horner et al. 1988) and aquatic environments (Stout 1989, Ostrofsky 1993, Campbell & Fuchshuber 1995). Tannin concentration thus could be an important indicator of chemical litter quality when addressing a variety of ecological questions relating to litter use and turnover.

The structural diversity of proanthocyanidins provides challenges for accurate quantitative analyses. Chromatographic characterization of cleavage products is therefore increasingly being used (Waterman & Mole 1994, Hernes & Hedges 2000), especially when specific functions of tannins are to be elucidated. Nevertheless, two simple methods for determining total proanthocyanidins, are considered to give ecologically meaningful information; these are known as the vanillin and the acid butanol assay, respectively (Hagerman & Butler 1989, Waterman & Mole 1994, Kraus et al. 2003b). Since oxidative cleavage of proanthocyanidins in alcohols yields anthocyanidins under strongly acidic conditions and the cleavage products (mainly cyanidin and delphinidin) absorb light in the visible range, proanthocyanidins can be quantified spectrophotometrically following depolymerization. The acid butanol assay recommended by Hagerman & Butler (1989) and Waterman & Mole (1994) for determining total proanthocyanidins is based on this reaction.

Table 16.1. Range of relative condensed tannin contents of undecomposed leaf litter from woody plant species.

Leaf material	Tannin concentration	Reference
5 *Acer* species	0.015—0.128[a]	1
6 *Quercus* species	0.017—0.107[a]	1
37 other woody plant species	0.003—0.276[a]	1
6 tropical *Eucalyptus* species	9—25[b]	2
6 nontropical *Eucalyptus* species	8—25[b]	2
6 non-*Eucalyptus* species	1—21[b]	2
4 *Populus* species or hybrids	0—53.3[c]	3

[a] Values are optical densities per mg of extracted dry leaf material; [b] Values are arbitrary relative numbers; [c] Values are given in mg g^{-1} leaf dry mass with tannin extracted from *Populus angustifolia* used as standard; 1 = Ostrofsky (1993); 2 = Campbell & Fuchshuber (1995); 3 = Driebe & Whitham (2000).

Before tannins can be analyzed, they need to be extracted from the sample matrix. Various extractants and extraction procedures have been described. Their relative efficiency depends on the analyzed material, due to differences in both tannin structure and the sample matrix (Waterman & Mole 1994, Yu & Dahlgren 2000), making compromises unavoidable when analyzing a range of different plant materials in comparative studies. One of the most common and frequently recommended extraction solvents is 50% methanol (Hagerman 1988, Waterman & Mole 1994); it is used in the procedure described below. The exact extraction procedure presented here has not been previously published, whereas the proposed protocol of the acid butanol assay has been adopted from Porter et al. (1986) and is also described in the comprehensive review by Waterman & Mole (1994).

2. EQUIPMENT, CHEMICALS AND SOLUTIONS

2.1. Equipment and Material

- Freeze-dryer
- Mill

- Analytical balance
- Glass tubes (10 ml, pressure resistant, with Teflon-lined screw-caps)
- Multiple-position magnetic stirrer (e.g. Variomag Telesystem HP 15 or Poly 15, or IKAMAG RO 15 Power, all with 15 stirring points)
- Disposable syringes (5 ml)
- Custom-made rack holding syringes upright on the magnetic stirrer
- Glass fibre filters (e.g. GF/F, Whatman)
- Cork borer (well sharpened; size matching the inner diameter of syringes)
- Stop cocks with Luer lock fitting the syringe tips
- Magnetic stirring bars (5 mm length)
- Volumetric flasks (100, 500, 1000 ml)
- Pipettes (e.g. Eppendorf Multipette and/or Varipette; 100—500 µl and 7 ml)
- Glass vials (e.g. 1.6-ml HPLC vials, with Teflon-lined caps), individually weighed to the nearest 0.1 mg
- Test tubes (10 ml)
- Vortex
- Water or dry bath (95 °C)
- Spectrophotometer (set at 550 nm)

2.2. Chemicals

- Methanol, reagent grade
- Deionized water (e.g. Nanopure®)
- $FeSO_4 \cdot 7 H_2O$
- n-Butanol, reagent grade
- Concentrated HCl (37%)
- Quebracho tannin (preferably purified; see http://www.users.muohio.edu/ hagermae), optional

2.3. Solutions

- Solution 1: 50% methanol:H_2O (v/v).
- Solution 2: Dissolve 700 mg $FeSO_4 \cdot 7 H_2O$ in 50 ml conc. HCl and adjust volume to 1000 ml with n-butanol.
- Solution 3: Stock solution of quebracho tannin standard (10—100 mg l^{-1}, depending on purity of tannin): weigh out 10—100 mg of (purified) quebracho tannin to the nearest 0.1 mg and dissolve in 100 ml of Solution 1, then dilute 10 fold with Solution 1.
- Standards: Use Solutions 1 (50% methanol) and 3 to prepare quebracho tannin standard solutions in the range 0—2.0 mg ml^{-1} or lower depending on purity of standard used.

3. EXPERIMENTAL PROCEDURES

3.1. Tannin Extraction

1. Dry leaves and grind to powder that passes through a 0.5-mm mesh screen.
2. Cut discs from glass fibre filters with a well-sharpened cork borer and place inside the disposable syringes.
3. Connect syringes to stop cocks with valves closed.
4. Add 50 mg sample material (weighed to the nearest 0.1 mg) to the syringes.
5. Place a small stirring bar in each syringe.
6. Place syringes on custom-built rack on the magnetic stirrer. The Luer ends of the syringes may have to be slightly shortened to minimize the vertical distance between the surface of the magnetic stirrer and the stirring bars in the syringes, so as to ensure continuous movement of the bars during extraction.
7. Add 400 µl of 50% methanol (Solution 1).
8. Connect plungers to the top of syringe barrels.
9. Extract tannins for 30 min with stirring at room temperature.
10. Filter extract directly into tared HPLC vials by slowly pushing plunger into the syringe barrel.
11. Repeat extraction three more times with 350 µl of Solution 1 (50% methanol) each time.
12. Rinse the stop cock with 50 µl methanol (50%) after the first two extraction steps.
13. Cap vials and reweigh them to the nearest 0.1 mg.
14. Calculate the volume of the extract, assuming a density of 0.9266 g ml^{-1} for 50% methanol.

3.2. Spectrophotometric Analysis

1. Pipette exact volume of 100—500 µl sample extract in test tube.
2. Add appropriate volume of deionized water to adjust total volume (i.e. sample extract plus water) to 500 µl.
3. Add 7 ml of Solution 2 ($FeSO_4 \cdot 7 H_2O$) and vortex.
4. Measure absorbance at 550 nm (control to correct for colour of extract).
5. Place tube in water bath at 95 °C and incubate for exactly 50 min.
6. Let cool to room temperature before measuring absorbance again at 550 nm.
7. Calculate absorbance due to the acid butanol reaction by subtracting the absorbance before heating from that after heating.
8. If (purified) quebracho tannin is available, proceed in the same way with the standard tannin solutions to establish a standard curve.
9. Express results in relative units or, preferably, in (purified) quebracho tannin equivalents based on absorbance readings and the standard curve.

4. FINAL REMARKS

Acetone interferes with the acid butanol assay. Consequently, the acetone-water mixtures commonly used for extracting tannins (e.g. Chapter 15) cannot be used unless the extract is completely evaporated and the residue redissolved in a solvent compatible with the assay.

The assay is very sensitive to varying amounts of water. It is essential to ensure, therefore, that the volumetric ratio of Solution 1 and 2 is always 1:14 (e.g. 500 μl of Solution 1 plus 7 ml of Solution 2). The water content is then 6.8%, which is close to the water content found by Porter et al. (1986) to yield the highest colour yield.

Waterman & Mole (1994) suggested not using an unheated reagent-sample mixture because some substances in plant tissue may yield red coloration even without heating. However, in our experience with a wide range of leaf litter from deciduous trees and shrubs, this potential problem is not generally encountered. Conversely, the substitution of HCl by H_2O as recommended by Waterman & Mole (1994) can result in precipitates.

A proanthocyanidin standard of sufficient purity is not commercially available, limiting quantitative comparisons among studies. To improve this situation, the use of purified quebracho tannin has been recommended; a protocol for purification – along with a wealth of useful information on tannin structural chemistry, other purification methods, biological activities and biosynthesis – can be downloaded from http://www.users.muohio.edu/hagermae, maintained by A.E. Hagerman. Alternatively, commercial cyanidin can be used as a relative standard (Hagerman & Butler 1989), keeping in mind that its colour yield differs from that of delphinidin. Procyanidin and prodelphinidin are also commercially available.

The standard curve may be discontinuous, the reason for which is unknown (Waterman & Mole 1994). One possibility to circumvent this effect may be to dilute sample extracts and use 5-cm or 10-cm cuvettes instead of standard 1-cm cuvettes for spectrophotometric measurements.

5. REFERENCES

Ayres, M.P., Clausen, T.P., MacLean, F. Jr., Redman, A.M. & Reichardt, P.B. (1997). Diversity of structure and antiherbivore activity in condensed tannins. *Ecology*, 78, 1696-1712.

Campbell, I.C. & Fuchshuber, L. (1995). Polyphenols, condensed tannins, and processing rates of tropical and temperate leaves in an Australian stream. *Journal of the North American Benthological Society*, 14, 174-182.

Driebe, E.M. & Whitham, T.G. (2000). Cottonwood hybridization affects tannin and nitrogen content of leaf litter and alters decomposition. *Oecologia*, 123, 99-107.

Hagerman, A.E. (1988). Extraction of tannin from fresh and preserved leaves. *Journal of Chemical Ecology*, 14, 453-461.

Hagerman, a. E. & Butler, L. G. 1989. Choosing appropriate methods and standards for assaying tannin. *Journal of Chemical Ecology*, 15, 1795–1810.

Hättenschwiler, S. & Vitousek, P.M. (2000). The role of polyphenols in terrestrial ecosystem nutrient cycling. *Trends in Ecology and Evolution*, 15, 238-243.

Hernes, P.J. & Hedges, J.I. (2000). Determination of condensed tannin monomers in environmental samples by capillary gas chromatography of acid depolymerization extracts. *Analytical Chemistry*, 72, 5115-5124.

Horner, J.D., Gosz, J.R. & Cates, R.G. (1988). The role of carbon-based plant secondary metabolites in decomposition in terrestrial ecosystems. *The American Naturalist*, 132, 869–883.

Kraus, T.E.C., Dahlgren, R.A. & Zasoski, R.J. (2003a). Tannins in nutrient dynamics of forest ecosystems – a review. *Plant and Soil*, 256, 41-66.

Kraus, T.E.C., Yu, Z., Preston, C.M.R., Dahlgren, A. & Zasoski, R.J. (2003b). Linking chemical reactivity and protein precipitation to structural characteristics of foliar tannins. *Journal of Chemical Ecology*, 29, 703-730.

Ostrofsky, M.L. (1993). Effect of tannins on leaf processing and conditioning rates in aquatic ecosystems: an empirical approach. *Canadian Journal of Fisheries and Aquatic Sciences*, 50, 1176-1180.

Ostrofsky, M.L. (1997). Relationship between chemical characteristics of autumn-shed leaves and aquatic processing rates. *Journal of the North American Benthological Society*, 16, 750-759.

Porter, L.J., Hrstich, L.N. & Chan, B.G. (1986). The conversion of procyanidin and delphinidin. *Phytochemistry*, 25, 223-230.

Stout, R.J. (1989). Effects of condensed tannins on leaf processing in mid-latitude and tropical streams: a theoretical approach. *Canadian Journal of Fisheries and Aquatic Sciences*, 46, 1097-1106.

Swain T. 1979. Tannins and lignins. In: G. Rosenthal & D.H. Janzen (eds.). *Herbivores: Their Interaction with Secondary Plant Metabolites*, pp. 657-682. Academic Press. New York.

Waterman, P.G. & Mole, S. (1994). *Analysis of Phenolic Plant Metabolites*. Blackwell Scientific Publications, Methods in Ecology. Oxford.

Yu, Z. & Dahlgren, R.A. (2000). Evaluation of methods for measuring polyphenols in conifer foliage. *Jounal of Chemical Ecology*, 26, 2119-2140.

Zimmer, M., Pennings, S.C., Buck, T.L. & Carefoot, T.H. (2002). Species-specific patterns of litter processing by terrestrial isopods (Isopoda: Oniscidea) in high intertidal salt marshes and coastal forests. *Functional Ecology*, 16, 596-607.

Zucker, W.V. (1983). Tannins: does structure determine function? An ecological perspective. *American Naturalist*, 121, 335-365.

CHAPTER 17

PROXIMATE LIGNIN AND CELLULOSE

MARK O. GESSNER

Department of Limnology, EAWAG, Limnological Research Center, 6047 Kastanienbaum, Switzerland.

1. INTRODUCTION

Lignin and cellulose are structural constituents of vascular plants that can make up a substantial part of litter dry mass (Table 17.1). Both compounds confer toughness to plant tissues (i.e., compressive and tensile strength; Chapter 18). Consequently, plant litter rich in these compounds tends to be highly refractory, with high concentrations particularly of lignin being conducive to slow litter decomposition (Gessner & Chauvet 1994, Berg & McClaugherty 2003).

On leaf litter decaying in streams, both biomass accumulation and sporulation activity of fungi decrease as litter lignin concentrations increase, suggesting that the negative effect of lignin is mediated at least partly through an impact on fungal decomposers (Gessner & Chauvet 1994, Maharning & Bärlocher 1996). In addition, lignin and cellulose concentrations may influence litter palatability to leaf-shredding invertebrates and hence litter consumption by these shredders. Freshwater invertebrates typically lack the enzymatic complements to digest cellulose and lignin; therefore, diets rich in these compounds are of poor nutritional quality to shredders, and this may have negative consequences for their survival, growth rate and fecundity (Bärlocher 1985, Suberkropp 1992, Graça 1993, Rong et al. 1995). However, some taxa (e.g. some *Tipula* species) may gain access to at least cellulose by means of a symbiotic cellulose-degrading gut flora (Kukor & Martin 1987, Martin 1987).

A variety of methods have been used to determine cellulose and lignin in plant tissues (e.g., McLellan et al. 1991, Van Soest et al. 1991, Hatfield et al. 1999). One simple approach, which has been widely used for forage fibre analyses and litter decomposition studies in both terrestrial and aquatic environments, consists of determining the residual weight of samples following successive removal of various tissue constituents. The first step is the extraction of components soluble in an acid detergent. Results by Ryan et al. (1990) suggest that with tree leaves and wood this approach produces similar results as the somewhat more complicated 'forest

M.A.S. Graça, F. Bärlocher & M.O. Gessner (eds.), Methods to Study Litter Decomposition: A Practical Guide, 115 – 120.

products analyses'. Since the approach does not necessarily determine concentrations of cellulose and lignin as defined chemically, the fractions resulting from the forage fibre method are referred to as proximate cellulose and lignin.

. The aim of the method presented here is to assess the concentrations of proximate lignin and cellulose in plant litter. Concentrations are determined gravimetrically using the acid-detergent fibre procedures proposed by Goering & Van Soest (1970) with slight modifications.

Table 17.1. Concentrations of proximate lignin (Gessner & Chauvet 1994) and cellulose (Gessner, unpubl. data) in undecomposed leaf litter as determined with the fibre forage method by Goering & Van Soest (1970). Values are means ± 1 SD.

Leaf species	Lignin (% leaf dry mass)	Cellulose (% leaf dry mass)
Fraxinus excelsior	6.8 ± 0.3	18.6 ± 1.0
Prunus avium	8.4 ± 1.0	16.3 ± 0.3
Alnus glutinosa	8.0 ± 0.7	13.8 ± 1.5
Corylus avellana	13.3 ± 0.9	23.3 ± 1.6
Platanus hybrida	30.9 ± 0.8	24.8 ± 1.0
Fagus sylvatica	25.5 ± 0.8	32.2 ± 2.8
Quercus ilex	18.5 ± 1.0	27.8 ± 3.0

2. EQUIPMENT, CHEMICALS AND SOLUTIONS

2.1. Equipment and Material

- Analytical balance
- Desiccator
- Dried sample powder ground to pass a 0.5-mm mesh-screen
- Eight screw-cap extraction tubes (approx. 40 ml, pressure-resistant)
- Dry bath or water bath (100 °C) with submersible rack holding at least 8 tubes
- Sixteen crucibles, Gooch type, porosity no. 2
- Filter manifold or individual units adapted for holding 8 crucibles (individual pressure regulation preferable)
- Pump for creating vacuum in filtration systems
- Hot plate or kettle for boiling H_2O
- Eight small trays (e.g. 10 × 15 cm) resistant to 72% sulphuric acid
- Latex gloves
- Eight acid-resistant spatulas or glass rods (about 8 cm long)
- Drying oven set at 105 °C
- Muffle furnace set at 550 °C

2.2. Chemicals

- Sulphuric acid, 0.5 M (reagent grade)
- Hexadecyltrimethylammonium bromide = Cetyltrimethylammonium bromide (CTAB), 20 g l^{-1}
- Decahydronaphtalene (reagent grade)
- Acetone (reagent grade) in spray bottles
- Sulphuric acid, 72% by weight (reagent grade)

2.3. Solutions

- Solution 1: Acid detergent solution: prepare 0.5 M sulphuric acid from low-molarity stock solution, check molarity by titration, adjust if necessary, then add the detergent CTAB (20 g l^{-1}) and stir. During handling of acid wear laboratory coat, security glasses and latex gloves.
- Solution 2: Prepare sulphuric acid at 72% by weight as described below. Weigh required amount of water into a volumetric flask and add the calculated amount of H$_2$SO$_4$ in *small* portions and *very slowly* with occasional swirling. Caution: heat production with risk of explosion hazard! Constantly cool flask in a water bath (e.g. sink). Allow sufficient time for cooling. Do not fill up flask to calibration mark. Finally let cool to 20 °C and adjust to exact volume. At all times during handling of acid wear laboratory coat, security glasses and latex gloves.
- Preparation of an acid solution: Given an acid at a concentration of A% and a density, δ, an acid at the concentration of X% is obtained as follows:
  - In mass units (for 100 g of acid solution):
    100 · (X/A) of acid at the concentration A%
    100—100 · (X/A) of H$_2$O
  - In volumetric units (e.g. in ml):
    100 · (X/A)/D of acid at the concentration A%
    100—100 · (X/A) of H$_2$O

  For example, for sulphuric acid at 72% starting with 96% ($\delta = 1.83$ g cm^{-3}):
  - For 100 g of solution:
    100 · (72/96) = 75.0 g of acid at 96%
    100—75.0 = 25.0 g of H$_2$O
  - Or in volume units:
    100 · (72/96)/1.83 = 41.0 ml of acid at 96%
    100—100 · (72/96) = 25.0 ml of H$_2$O

3. EXPERIMENTAL PROCEDURES

3.1. Sample Preparation

1. Weigh clean and oven-dry crucibles to the nearest 0.1 mg.

2. Weigh air-dry sample ground to pass a 0.2 mm-mesh screen (245—255 mg to the nearest 0.1 mg) and place in extraction tube.
3. Weigh same amount of sample in ignited, tared porcelain or aluminium pans for determining moisture content and ash-free dry mass.
4. Add to tubes 20 ml of acid-detergent solution and 0.4 ml decahydronaphatalene.

3.2. Acid-Detergent Fibre Determination

1. Heat tubes to boiling for 5—10 min in a water bath with occasional swirling.
2. Reduce heat as boiling begins to avoid foaming. Boil for 60 min from onset of boiling. Adjust boiling to a slow, even level.
3. Filter tube content on a tared Gooch crucible set on a filter manifold. Use light suction! Recover particles in tubes quantitatively. Break up the filtered mat with a spatula or glass rod and wash twice with hot water (90—100 °C). Rinse sides of the crucible in the same manner.
4. Repeat wash with acetone until it removes no more colour. Break up all lumps so that the solvent comes into contact with all particles of fibre.
5. Suck the acid-detergent fibre free of acetone and dry overnight at 105 °C.
6. Place oven-dry crucible in desiccator for 1 h and then weigh to nearest 0.1 mg.
7. Calculate acid-detergent fibre (ADF) as follows:

$$\frac{W_0 - W_t}{W_S} \cdot 100 = ADF \qquad (17.1)$$

where: W_0 = weight of the oven-dry crucible including fibre
W_t = tared weight of the oven-dry crucible
W_S = oven-dry sample weight.

8. Correct value for moisture content of sample.

3.3. Acid-Detergent Lignin and Cellulose Determination

9. Cover the contents of the crucible with cooled (15 °C; water bath) 72% H_2SO_4 and stir with a spatula or glass rod to a smooth paste breaking all lumps.
10. Fill crucible about half with acid and stir. Let spatula or glass rod remain in crucible.
11. Refill with 72% H_2SO_4 and stir at hourly intervals as acid drains away. Crucibles do not need to be kept full at all times, but samples must be covered continuously. Three additions of acid suffice. Keep crucible at 20—23 °C.
12. Filter off after 3 h as much acid as possible with vacuum (start with weak vacuum).
13. Wash contents abundantly with hot water until free from acid. Rinse and remove stirring rod.
14. Dry crucible overnight at 105 °C.

15. Place crucible in desiccator for 1 h and weigh to the nearest 0.1 mg.
16. Ignite crucible in a muffle furnace at 550 °C for 3 h and then cool to 105 °C.
17. Place in desiccator for 1 h and weigh.
18. Calculate acid-detergent cellulose (ADC) as follows:

$$\frac{L_a}{W_S} \cdot 100 = ADC \qquad (17.2)$$

where: L_a = loss due to 72% H_2SO_4 treatment
W_S = oven-dry sample weight.
19. Calculate acid-detergent lignin as follows:

$$\frac{L_i}{W_S} \cdot 100 = ADL \qquad (17.3)$$

where: L_i = loss upon ignition after 72% H_2SO_4 treatment
W_S = oven-dry sample weight.
20. Correct values for moisture content of sample.

4. REFERENCES

Bärlocher, F. (1985). The role of fungi in the nutrition of stream invertebrates. *Botanical Journal of the Linnean Society*, 91, 83-94.

Berg, B. & McClaugherty, C. (2003). *Plant Litter – Decomposition, Humus Formation, Carbon Sequestration*. Springer. Berlin.

Gessner M.O. & E. Chauvet. (1994). Importance of stream microfungi in controlling breakdown rates of leaf litter. *Ecology*, 75, 1807-1817.

Goering, H.K., & Van Soest, P.J. (1970). Forage fiber analyses (apparatus, reagents, procedures, and some applications). *Agriculture Handbook 379* (pp. 1-20). U.S. Department of Agriculture. Washington DC.

Graça, M.A.S. (1993). Patterns and processes in detritus-based stream systems. *Limnologica*, 23, 107-114.

Hatfield, R.D., Grabber, J., Ralph J., & Brei K. (1999). Using the acetyl bromide assay to determine lignin concentrations in herbaceous plants: some cautionary notes. *Journal of Agricultural and Food Chemistry*, 47, 628-632.

Kukor, J.J., & Martin, M.M. (1987). Nutritional ecology of fungus-feeding arthropods. In: F. Slansky Jr. & J.G. Rodriguez (eds.), *Nutritional Ecology of Insects, Mites, Spiders, and Related Invertebrates* (pp. 791-814). John Wiley & Sons. New York.

Maharning, A.R., & Bärlocher, F. (1996). Growth and reproduction in aquatic hyphomycetes. *Mycologia*, 88, 80-88.

Martin, M.M. (1987). Acquired enzymes in detritivores. In: M.M. Martin (ed.), *Invertebrate-Microbial Interactions. Ingested Fungal Enzymes in Arthropod Biology* (pp. 49-72). Comstock Publishing Associates. Ithaca.

McLellan, T.M., Aber, J.D., & Martin, M.E. (1991). Determination of nitrogen, lignin and cellulose content of decomposing leaf material by near infrared reflectance spectroscopy. *Canadian Journal of Forest Research*, 21, 1684-1688.

Rong, Q., Sridhar, K.R., & Bärlocher, F. (1995). Food selection in three leaf-shredding stream invertebrates. *Hydrobiologia*, 316, 173-181.

Ryan, M.G., Melillo, J.M., & Ricca, A. (1990). A comparison of methods for determining proximate carbon fractions of forest litter. *Canadian Journal of Forest Reserch*, 20, 166-171.

Suberkropp, K. (1992). Interactions with invertebrates. In: F. Bärlocher (ed.), *The Ecology of Aquatic Hyphomycetes, Ecological Studies, Vol. 94* (pp. 118-133). Springer. Berlin.

Van Soest, P.J., Robertson, J.B., & Lewis, B.A. (1991). Carbohydrate methodology, metabolism, and nutritional implications in dairy cattle. *Journal of Dairy Science*, 74, 3583-3597.

CHAPTER 18

LEAF TOUGHNESS

MANUEL A.S. GRAÇA[1] & MARTIN ZIMMER[2]

[1]*Departmento de Zoologia, Universidade de Coimbra, 3004-517 Coimbra, Portugal;* [2]*Zoologisches Institut: Limnologie, Christian-Albrechts-Universität Kiel, 24098 Kiel, Germany.*

1. INTRODUCTION

Studies on plant decomposition have emphasized the role of internal (plant characteristics) and external (environment) factors in the decomposition of leaves and their consumption by detritivores (e.g. Gallardo & Merino 1993, Zimmer & Topp 1997, 2000, Gessner et al. 1997). An important internal leaf factor is chemistry, particularly concentrations of nutrients such as nitrogen (Chapter 8), structural compounds particularly lignin and cellulose (Chapter 17), and plant chemical defences such as polyphenolics (Chapters 14 and 15). Internal factors also include physical leaf attributes such as waxy cuticles and particularly leaf toughness.

Toughness of plant tissues impede feeding of terrestrial (Cornelissen et al. 1999), marine (Pennings & Paul 1992) and freshwater (Arsuffi and Suberkropp 1984) invertebrates. Initial toughness was a significant predictor of leaf breakdown rates of 10 species in a decomposition experiment in New Zealand streams (Quinn et al. 2000). In the same experiment, leaf toughness after 2 days of leaching was also correlated with nutrient uptake by leaf-colonizing microbes. Leaf toughness is therefore one of the main factors affecting invertebrate feeding, microbial decomposition and susceptibility of leaves to physical fragmentation.

Leaf toughness correlates with fiber and lignin (Chapter 17) and can be estimated by measuring the force needed to penetrate a leaf sample (e.g. Arsuffi and Suberkropp 1984), or the force needed to tear apart a leaf sample (Graça et al. 1993). These two leaf properties of resistance to physical disruption, fissure and tear, are controlled by different chemical constituents in intact leaves, lignin and cellulose fibers, respectively, and are not necessarily related. Penetrometers are commercially available but are relatively expensive. Here we describe two devices that can be built in the laboratory with simple, inexpensive materials. Both have been used successfully to measure the toughness of leaves decomposing in streams (Graça et

M.A.S. Graça, F. Bärlocher & M.O. Gessner (eds.), Methods to Study Litter Decomposition: A Practical Guide, 121 – 126.

al. 1993), a salt marsh (Graça et al. 2000) and soil (Zimmer and Topp 1997, 2000), and provide complementary information on leaf resistance to physical disruption.

2. PENETROMETER

2.1. Equipment

The principle of measuring the force of leaf penetration was described by Williams (1954), and later modified by Tanton (1962). A leaf disc is clamped firmly into the base of a penetrometer (Fig. 18.1) consisting of a basal plate (ba) and a fixing device (fi), with a central hole, the latter held in space by screws. The fixing device should be made of an acrylic tile in order to allow checking the position of the leaf disc to make sure that the punch (pu) fitting the central hole of the base does not get into contact with any high-order vein of the leaf. Ideally, the punch is made of a metal

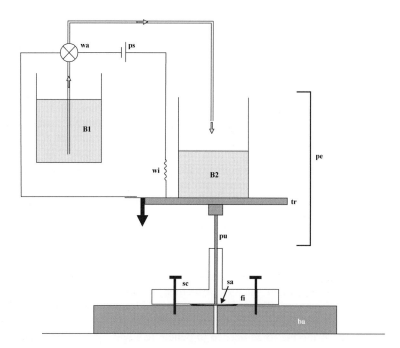

Figure 18.1. Schematic view of a device to measure resistance of leaf material against penetrating forces. ba = base of penetrometer with central hole B1 = beaker 1 (water reservoir), B2 = beaker 2 collecting water pumped from the reservoir, fi = fixing device with central hole, held in place by screws (sc), pe = penetrator (punch + tray + cup 2 + pumped water), ps = 9V DC power source, pu = punch, penetrating leaf disc sample, sa = leaf disc sample, sc = screw fixing the leaf disc sample, tr = tray for beaker, wa = water pump, wi = flexible wire.

rod (diameter 2 mm) with a rounded tip. The top of the punch is fixed to a tray (tr) to carry a beaker (C2) that will be filled with water to increase the mass of the penetrator (= punch + tray + beaker + water) until a critical value is reached and the punch cuts through the leaf.

Penetrometers can be operated manually or semi-automatically. Manual operation requires pouring water into a beaker (B2) by hand until the critical mass is reached. With a semi-automated instrument (Fig. 18.1), water is pumped from a reservoir (B1) into the beaker (B2) by an electric pump (wa) until the circuit is interrupted when the punch penetrates the leaf disc and the penetrator abruptly changes its position.

2.2. Procedure

1. Cut discs (8—12 mm diameter) from a leaf using a cork borer. Avoid main veins.
2. Attach a disc to the penetrometer base such that no high-order vein impedes leaf penetration by the punch.
3. Gradually add water to the beaker of the penetrator until the punch penetrates the leaf disc. Beakers of several sizes are useful since leaf toughness may vary greatly among species.
4. Weigh the water-filled beaker. The mass of the penetrator (punch + tray + beaker + water) is proportional to leaf toughness.
5. Leaf toughness can be expressed as the critical mass (g) of a penetrator with a standard punch needed to penetrate a leaf (e.g. Zimmer & Topp 2000). Alternatively, mass units can be transformed into penetration pressure (kPa) = mass (g) × gravity (e.g. 9.807 m s^{-2} at 45° latitude) ÷ area of the rod penetrating the leaf (mm^2) (Quinn et al. 2000).

3. TEARING DEVICE

3.1. Equipment

This device is composed of two pegs securing a leaf disc (Fig. 18.2). Pegs can be made from wooden clothes-pegs with the anterior end cut out. One of the pegs is fixed to a ring stand by a clamp. The other peg is connected to a cup via a string passing through a pulley. To measure the resistance to tearing, a sample is secured between the two pegs. Sand is then gradually added to the cup via a funnel until the mass of the accumulated sand exerts a force that tears apart the leaf. The mass needed to reach this point is proportional to leaf toughness.

3.2. Procedure

1. Cut discs (8—12 mm diameter) from a leaf using a cork borer. Avoid main veins.
2. Secure the disc between the two pegs. Be consistent regarding the position of the sample veins, i.e. position the veins parallel to the pegs.
3. Gradually add sand to the cup until the leaf disc breaks into two. Cups of several sizes are useful since leaf toughness may vary greatly among leaf samples.
4. Weigh the cup. The mass of the cup is proportional to leaf tear strength, which can also be expressed as a force (kN), analogous to the penetration pressure indicated above.

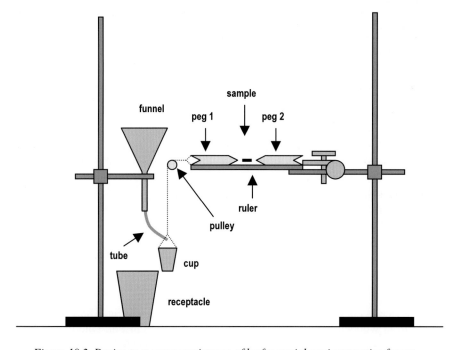

Figure 18.2. Device to measure resistance of leaf material against tractive forces.

4. REFERENCES

Arsuffi, T. L. & Suberkropp, K. (1984). Leaf processing capabilities of aquatic hyphomycetes: interspecific differences and influence on shredder feeding preferences. *Oikos*, 42, 144-154.

Cornelissen, J.H.C., Perez-Harguindeguy, N., Diaz, S., Grime, J.P, Marzano, B., Cabido, M., Vendramini, F. & Cerabolini, B. (1999). Leaf structure and defence control litter decomposition rate across species and life forms in regional floras on two continents. *New Phytologist*, 143, 191-200.

Gallardo, A. & Merino, J. (1992). Nitrogen immobilization in leaf litter at two Mediterranean ecosystems of Spain: influence of substrate quality. *Ecology*, 74, 152-161.

Gessner, M.O., Suberkropp, K. & Chauvet E. (1997). Decomposition of plant litter in marine and freshwater ecosystems. In D.T. Wicklow & B. Söderström (eds.). *The Mycota, Vol. IV: Environmental and Microbial Relationships.* Springer-Verlag, Berlin.

Graça, M.A.S., Newell, S.Y. & Kneib, R.T. (2000). Consumption rates of organic matter and fungal mass of the *Spartina alterniflora* decay system by three species of saltmarsh invertebrates. *Marine Biology*, 136, 281-289

Graça, M.A.S., Maltby, L. & Calow, P. (1993). Importance of fungi in the diet of *Gammarus pulex* and *Asellus aquaticus*. *Oecologia*, 96, 304-309.

Pennings, S.C. & Paul, V.J. (1992). Effect of plant toughness, calcification, and chemistry on hervibory by *Dolabella auricularia*. *Ecology*, 73, 1606-1619.

Quinn, J.M., Burrel, G.P. & Parkyn, S.M. (2000). Influences of leaf toughness and nitrogen content on in-stream processing and nutrient uptake by litter in a Waikato, New Zealand, pasture stream and streamside channels. *New Zealand Journal of Marine and Freshwater Research*, 34, 253-271.

Tanton, M.T. (1962). The effect of leaf "toughness" on the feeding of larvae of the Mustard beetle *Phaedoncochleariae* Fab. *Entomologia Experimentalis et Applicata*, 5,74-78.

Williams, L.H. (1954). The feeding habits and food preference of Acrididae and the factors which determine them. *Transactions of The Royal Entomological Society of London*, 105, 423-545.

Zimmer, M. & Topp, W. (1997). Does leaf litter quality influence population parameters of the common woodlouse, *Porcellio scaber* Latr., 1804 (Crustacea: Isopoda)? *Biology and Fertility of Soils*, 24,435-441.

Zimmer, M. & Topp, W. (2000). Species-specific utilization of food sources by sympatric woodlice (Isopoda: Oniscidea). *Journal of Animal Ecology*, 69, 1071-1082.

PART 3

MICROBIAL DECOMPOSERS

CHAPTER 19

TECHNIQUES FOR HANDLING INGOLDIAN FUNGI

ENRIQUE DESCALS

Instituto Mediterráneo de Estudios Avanzados (CSIC/Univ. Illes Balears). Calle Maestro Miquel Marquès, 21, 07190 Esporles, Mallorca, Balears, Spain.

1. INTRODUCTION

Numerous taxonomic surveys and ecological studies on Ingoldian fungi (also known as "aquatic hyphomycetes") have been published since Ingold's seminal paper of 1942 (Bärlocher 1992). Either conidiophores (Shearer & Webster 1985) or, more commonly, detached conidia, are used for identification and in some cases quantification (Gessner et al. 2003; Chapter 24). Identification is facilitated by the often characteristic conidial shapes (see Chapter 21), which are uncommon among fungi in general. Even so, unknown and doubtful conidial forms often comprise a third of the taxa listed in detailed surveys (e.g. Descals 1987, 1998), indicating that Ingoldian fungi remain insufficiently described despite decades of taxonomic and ecological work.

   Natural foam is the main source of Ingoldian fungi for biodiversity studies and conidial isolation into pure culture. Lather-like accumulations attached to boulders and other obstacles in small pools in streams can be a rich source. In hard waters, natural foam may not accumulate readily and will then contain few conidia. Muddy samples tend to be heavily contaminated with bacteria or yeasts. Alternative sources for sampling Ingoldian fungi are submerged plant litter, particularly decomposing leaves in or near streams, and stream water samples, which may contain up to 30000 conidia per litre (Webster & Descals 1981). Finally, conidia can accumulate in rain water throughfall from riparian canopies (and be collected in a funnel), from tree stem flow, dewdrops beaten off canopies (collected onto an inverted clean umbrella) and waterfall mist onto glass slides (Ando & Tubaki 1984).

   Ingoldian fungi typically form conidia under water, although some also sporulate in Petri dishes without colony submersion when relative humidity is high (e.g. *Calcarispora hiemalis*). Others sporulate both in air and water, but conidia in

M.A.S. Graça, F. Bärlocher & M.O. Gessner (eds.), Methods to Study Litter Decomposition: A Practical Guide, 129 – 142.

air often lack branches. The remaining species require contact with liquid water, after typically prolonged incubation (one to several days) in submerged or partly submerged conditions. In spite of the fully aquatic life cycle of the anamorphs, Ingoldian fungi with sexual reproduction are amphibious in that their teleomorphs must release their spores outside water. Meiospores (asco- and basidiospores), which unlike the conidia of anamorphs are not species-diagnostic, are therefore dispersed in air or on the water surface. An exception is *Loramyces juncicola*, a freshwater ascomycete with sigmoid ascospores (Ingold & Chapman 1952). It does not occur in streams, however. Only leptosphaeriaceous ascospores, probably belonging to *Massarina* species with Ingoldian anamorphs, are regularly recognized in stream drift, but have never been isolated.

Teleomorphs are so far known for only a minority of species. However, only a few taxonomists have searched for them, mostly in temperate climates. The few known cases are spread over a wide range of taxa (including Helotiales, Pezizales, nectriaceous forms, bitunicate fungi and corticiaceous and other basidiomycetes). It is therefore likely that many more remain to be discovered. Ascomycetous teleomorphs (pseudothecia and especially apothecia) are commonly found on wood (Willoughby & Archer 1973) collected along riverbanks, especially in the summer, and after moist incubation for several weeks. Teleomorph characters often allow the identification of Ingoldian species. For example, records of *Miladina lechithina* have been included in distributional studies of *Actinosporella megalospora* by Descals & Rodríguez (2002). Basidiomycetous teleomorphs are inconspicuous, filmy or gelatinous forms known only from pure culture.

Synanamorph conidia are occasionally observed in cultures in addition to the normal characteristically shaped conidia of Ingoldian fungi (see Chapter 21). Some of these synanamorph conidia can even be species-diagnostic (e.g. *Culicidospora gravida*), although most are small and inconspicuous. They tend to appear after slightly longer submerged or semisubmerged incubation and cannot be genetically associated to the Ingoldian anamorph except when seen in pure culture. Synanamorph conidia can be functionally spermatial. Inducing synanamorph formation is therefore a critical step for sexual reproduction and hence in the establishment of anamorph-teleomorph relationships.

Apart from synanamorph production and establishment of anamorph-teleomorph relationships, pure cultures of Ingoldian fungi are needed in descriptive work for recording secondary diagnostic characters such as colony pigmentation, morphology, and the presence of sclerotia and microsclerotia. They are also useful for confirming some identifications solely based on conidia; for increasing material for identification when growth and sporulation are sparse on the natural substrates; and for establishing culture collections for experimental, commercial or exchange purposes.

The present chapter deals with some common techniques and approaches to handle Ingoldian fungi. Techniques include analysis of conidia and/or conidiophores in foam, plant litter and water samples; isolation into pure culture; growth on agar plates and sporulation; and teleomorph induction. More detailed accounts are given in Descals (1997) and Descals & Moralejo (2001), and a variety of methods commonly used in ecological work are summarized in Gessner et al. (2003). Since correct species identification is often fundamental for ecological work, many techniques presented here emphasize preparation of fungal material for this purpose, complementing the taxonomic key on common species from temperate regions in Chapter 21. Standard techniques to isolate fungi present as mycelia, such as particle plating of leaf, wood or root tissues (e.g., Bärlocher & Kendrick 1974, Kirby 1987, Fisher & Petrini 1989), are not included in this chapter.

2. EQUIPMENT AND CHEMICALS

2.1. Equipment and Material

- Compound microscope with mechanical stage and 10, 20 and 40× objectives; intermediate magnification lenses (1.25—2×); built-in transformer for field work.; phase and differential interference contrast optics for microphotography, as aids for identification. Digital camera and drawing tube are needed for taxonomic (descriptive) purposes
- Inverted microscope for isolation (optional)
- Dissecting microscope (at least up to 100×) with transmitted light. Incident light will be needed for teleomorph detection or isolation
- Labelled jars and spoon for collecting foam in the field
- Polythene bags for collecting colonized leaves and other substrates as well as teleomorphs
- Chisel, mallet and folding handsaw for sampling colonized wood
- Bunsen burner with gas canisters (or alcohol lamp) for isolation
- Fan heater or other source of heat for fast drying of foam slide preparations
- Microtechnique kit: watchmaker's forceps, micro-scalpel consisting of mounted fine sewing needles or 00 insect pins with tip flattened hot with a hammer, surgical scalpel, wide loop for sampling foam for slide preparations, mounted hair, finder slides
- Filtration equipment for ecological work
- Cellulose-ester membrane filters
- Colour transparency or negative films, or digital camera
- Semi-transparent paper envelopes (such as used for storing photographic negatives) for storage of dried cultures (optional)
- Plankton sedimentation chambers or table centrifuge (optional)
- Tracing satin and India ink pens (for taxonomic work)
- General glassware, including small vials for specimen storage, sterile pipettes (10 and 1 ml), flame-pulled Pasteur pipettes, conical flasks
- Microscope slides (grease-free), coverslips (20 × 20 mm), slide boxes, slide labels
- Sterile disposable syringes for mass transfers of mycelia

- Aquarium aeration systems or periodic water renewal chambers (Suberkropp 1991) for inducing conidial production
- Near-UV and daylight lamps for anamorph and/or teleomorph induction
- Autoclave

2.2. Chemicals

- Ethanol (96%)
- Lugol's iodine fixative
- Lactofuchsin: acid fuchsin 0.1 g, lactic acid 100 ml (Kirk et al. 2001)
- Nail varnish (preferably containing nylon) or liquid cover glass such as Merckoglas®
- Dilute bleach (5% NaOCl)
- Sterile distilled water
- Broad-spectrum insecticide and acaricide, preferably containing an ovicide, available in agricultural or gardening supply shops
- Wetting agent (e.g. Tween-80)
- Antibiotics: chloramphenicol (up to 1 g l^{-1}, may be added before autoclaving the medium) or penicillin (1 million IU l^{-1}) plus streptomycin sulphate (1 g l^{-1}) added after autoclaving

2.3. Media

- Stock medium for colony pigment expression: 2% malt extract in 2% agar (2% MEA)
- Sporulation medium: 0.1% malt extract in 2% agar (0.1% MEA)
- Isolation medium: sporulation medium + antibiotics (see above) . These concentrations of antibiotics work well in most situations. However, the best concentration depends on the degree of contamination of the source material. Chloramphenicol controls bacterial growth even in rather dirty foam samples, but may suppress germination in some species of Ingoldian fungi (Gessner et al. 2003). Preparing plates a few days in advance allows discarding any contaminated plates. Another advantage is that the agar dries up slightly so that water drops used in single-sporing techniques (see Section 3.4 below) are rapidly adsorbed.

3. EXPERIMENTAL PROCEDURES

3.1. Foam Samples

1. Collect stream foam with a spoon and transfer to a jar. Decant excess water. If foam is scanty or breaks up quickly, collect subsamples in separate jars.

2. Preserve samples with a few drops of ethanol if isolation of cultures is not intended. Jars with foam to be used for isolating individual large conidia must be placed on melting ice and processed (see Sections 3.4 and 3.5) in less than 2 h.
3. With a Pasteur pipette, transfer 3—4 drops of liquefied foam onto a grease-free slide.
4. Air-dry drop, remove any sand grains with a needle, add a small drop of fixative, burst the odd gas bubble with a heated needle, cover, seal, label and store in slide box.
5. View preparations under a microscope at 100—500×, preferably with phase contrast optics.
6. Produce line drawings or microphotographs of conidia of doubtful or unknown identity.
7. Save semi-permanent lactofuchsin slide preparations for future identifications.
8. Optionally, obtain pure cultures as described in Sections 3.4 and 3.5 below.

3.2. Plant Litter Samples

1. Collect decaying plant litter such as leaves in the field and store in polythene bags.
2. Keep samples refrigerated and moist, but not flooded, for transport to the laboratory. Process samples promptly (i.e. within a few days at most) to avoid sporulation by fungal air contaminants.
3. Rinse the plant litter gently to avoid loss of delicate surface mycelia. Use sterile water. Tap water may be used if chlorine toxicity and contamination by Pythiaceae, terrestrial hyphomycetes, coelomycetes and other fungi is not a concern.
4. Observe fungal conidia and conidiophores on leaves (Fig. 19.1) directly under a dissecting microscope with high magnification (i.e. 200× or at least 100×). Or observe small pieces of leaf mounted in water on slides or hanging drops, and placed under the compound microscope. Fungal structures are best seen along the margins of litter pieces.

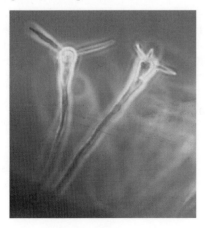

Figure 19.1. Clavariopsis aquatica *conidia attached to conidiophores on a maple leaf. Photo F. Bärlocher.*

5. For quantifying sporulation, submerge substrates under turbulent conditions (forced aeration or shaker) at stream or laboratory temperature for 1—3 or more days, renewing the water if necessary (Chapter 24, Gessner et al. 2003).
6. Harvest newly detached conidia by membrane filtration (see Chapter 24). Avoid letting conidial suspensions sit even for short periods, because conidial mucilage soon adheres firmly to any solid substrate, such as the walls of containers. If live conidia are not needed (e.g. for isolation), the stickiness can be neutralized by adding 10% KOH or formaldehyde (2% final concentration). Stirring suspensions may also help.
7. Render filter translucent by first staining in a water-soluble cytoplasmic stain, such as Waterman's ink diluted 10fold, air-drying and then flooding it in immersion oil.
8. Observe a section of filter at 100—200× with a compound microscope.
9. Produce drawings, microphotographs, slide preparations and/or pure cultures as described in sections above and below.

3.3. Water Samples

1. Collect stream water as a source of conidia if their concentration is expected to be high.
2. Filter the water through a membrane filter (typically 200—1000 ml), turn filter translucent (optional), and scan under a compound microscope at 100—500× (see Section 3.2, Chapter 24, Gessner et al. 2003).
3. If conidia are rarer, concentrate them in artificial foam. The technique has not been worked out, but try the following: add a drop of dilute Tween-80 (and a few drops of fixative if the sample is not intended for isolation) to 10 l of stream water, and decant the water back and forth from one bucket into another. Then collect and process foam samples as described in Section 3.1.
4. Produce drawings, microphotographs, slide preparations and/or pure cultures as described in sections above and below.

3.4. Isolation into Pure Culture

1. Prepare Petri dishes with an isolation medium.
2. For individual spore lifting techniques, draw a ladder or spoke pattern (Descals 1997) with indelible ink on the bottom lid of the isolation dish (Fig. 19.2).
3. Optionally, bore tiny wells with a hot needle in the centre of the squares or sectors, where the conidium will be placed. This traps most bacterial or yeast colonies and allows hyphae to grow out into clean agar.
4. Place some leaves collected in the field in Petri dishes with sterilized stream or distilled water in a Petri dish for 1—3 days. Dishes may be aerated or shaken in conical flasks to stimulate sporulation.

5. Lift large conidia (e.g. >400 μm in span) under a dissecting microscope with a mounted hair (if floating) or Pasteur pipette (if settled or in suspension) and place on the agar.
6. Since germination success of floating conidia is low, collect a large loopful from the agar surface containing several conidia and spread it over the medium.

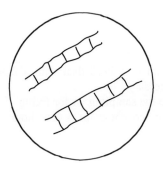

Figure 19.2.Ladder pattern drawn in the bottom lid of a Petri dish. Individual conidia are placed in the central cells and the germlings transferred to the neighbouring outer cells.

7. Isolate germlings after at least a few hours of incubation.
8. Conidia suspended in water or recently settled can be made to flow into a capillary tube (e.g., a Pasteur pipette pulled over the flame and covered with a perforated teat). Cover this hole with the thumb and press the teat gently to force the conidium out onto the agar. To sterilize the tip of the pipette, draw some water into it and place over the flame somewhat above the tip (the tip must not be flamed, as it will melt and become sealed)
9. After 24—48 h, check isolated conidia for germ tubes.
10. With a flamed microscalpel remove contaminant germlings lying nearby and cut out the section of agar containing the germling and transfer onto fresh medium.

3.5. Isolation with a Finder Slide

1. Isolate conidia too small to be handled under a dissecting microscope with the aid of a finder slide (Graticules Ltd., Tonbridge, UK; Gessner et al. 2003).
2. A finder can also be improvised by printing a letter-size template with the computer, as indicated in Fig. 19.3. Photograph with a colour transparency film. Cut out a frame and attach to microscope slide with nail varnish. Glue coverslip on top of the frame with more nail varnish. Put some weight on the slip and allow to dry overnight.

aa	ab	ac	ad	ae ...
ba	bb	bc	bd	be ...
ca	cb	cc	cd	ce ...
...	...	...	...	...

Fig. 19.3. Template printed on letter-size paper, to be photographed with a colour transparency film and placed under a microscope slide to locate conidia.

3. In the field, flood the isolation medium in a Petri dish with a foam sample. Dilute with sterile distilled water if necessary.
4. Decant the excess suspension back into the sampling jar for fixing.
5. Incubate the plates horizontally in an ice chest for a few hours to initiate germination.
6. Transfer a rectangular portion of the isolation medium onto the finder slide
7. Scan the surface under the compound microscope at 100×, check at 200× for contaminating spores attached to or lying in the vicinity of the conidium of interest. Record the coordinates under which lie the desired germlings.
8. Transfer the slide to the dissecting microcope, locate the coordinates, and, with a flamed microscalpel, remove nearby contaminant germlings and cut out the tiny section of agar containing the germling and transfer onto fresh isolation medium.

3.6. Colony Growth on Agar Plates

1. Let colonies grow under standard conditions to stimulate pigment production and induce colony characters needed for comparisons with published descriptions. 2% MEA is one of the more popular media used for describing fungal cultures.
2. If large amounts of mycelium are rapidly needed for experimental work, force a piece of agar colony through a sterile syringe (without the needle) onto agar medium. Add some sterile distilled water and spread the suspension evenly over the agar surface with a flamed bent glass rod. This results in rapid colony growth as a single carpet (intra-specific vegetative incompatibility between incipient colonies occasionally observed, prevents uniform occupation of the agar surface).
3. Seal cultures growing in Petri dishes with tape to reduce contamination by aerial spores and to limit medium dehydration.
4. To protect cultures from fungal and other contaminations caused by mite intrusion, spray bench surfaces and plates with an acaricide and swab working surfaces with a mineral oil. Sealing dishes with tape is not effective against mites.

3.7. Induction of Anamorph Sporulation

1. Place leaf material bearing fungi or fungi growing on agar medium in conical flasks with sterile water to induce anamorph formation. Use colonies grown on 0.1% MEA, since rich media such as 2% MEA may discourage sporulation. Four types of liquid can be used (Descals & Moralejo 2001): (1) sterile distilled water; (2) sterile filtered stream water (chemically variable, but may yield high conidium production; (3) tap water (chlorine may be inhibitory); and (4) dilute mineral solutions.
2. Incubate submerged agar or litter pieces for 1—2 days with forced aeration or shaking.
3. View preparations under a microscope at 100—500× and produce line drawings, microphotographs, and semi-permanent lactofuchsin slide preparations for documentation.

3.8. Teleomorph Induction

1. Add some water to pure agar cultures or field-collected material.
2. Agitate periodically to facilitate dispersing spermatia over the mycelia.
3. Supplement diffuse daylight with near UV to favour teleomorph induction.
4. Regularly search plate or field material over several months for formation of sexual structures.
5. Identify species using pertinent mycological literature. For world species, the overviews by Webster & Descals (1981) and Descals & Marvanová (ined.) are recommended; for temperate species, Ingold (1975) and Descals & Marvanová (ined.), and Chapter 21. Nawawi (1985) and Marvanová (1997) are particularly useful for tropical species. Consultation of original descriptions is recommended.

4. FINAL REMARKS

Membrane filtration of stream water (Section 3.3, Iqbal & Webster 1973, Gessner et al. 2003) has become widely used by ecologists. Potential disadvantages include loss of optical resolution due to incorrect staining, opaque background and/or mangling due to excessive vacuum. Techniques for concentrating conidia other than membrane filtration and production of artificial foam (see Section 3.3) may include (1) sedimentation by gravity overnight or for appropriate shorter periods (Chamier & Dixon 1982), (2) evaporation by any combination of vacuum, heat or ventilation; and (3) centrifugation. The last two approaches have not yet been tested.

Aerial conidia of Ingoldian fungi forming on damp-incubated substrates are sticky and cannot be easily lifted individually. They may be first transferred to a drop of sterile water on the isolation medium, spread with a small, sterile loop, allowed to germinate and isolated as in Sections 3.4. and 3.5.

Consider the following points when producing mycelia and/or spores from field-collected plant litter or pure cultures on agar:

- Turbulence: Although aeration and shaking stimulate sporulation, excess turbulence may cause fragmentation of complex conidia (e.g. *Dendrospora* and *Varicosporium*) and provoke bubble burst which can propel conidia out of the suspension.
- Incubation time: Incubations longer than the 1—2 days recommended in Section 3.7 may be needed to observe conidiogenesis and induce synanamorph and teleomorph formation. However, long incubations increase the risk of (a) conidial malformations which interfere with identification; (b) conidial repetition or microcycle conidiation which may artificially increase conidial counts (e.g. in *Articulospora tetracladia*, *Lemonniera* spp.). Conidia adhering to walls through mucilage production can be a problem even during shorter incubations.
- Temperature and nutrient concentrations: The optimum sporulation temperatures of Ingoldian fungi are usually well above those found in the source streams (e.g. Chauvet & Suberkropp 1998). Such temperatures may therefore be chosen during laboratory incubations (e.g. 10—20 °C), unless sporulation activities under natural conditions are to be estimated. Drastic changes in temperature and nutrient concentrations should be avoided, however, because they may permanently disrupt sporulation.
- Light: Pulses or protracted treatments with near UV light may stimulate growth (e.g. *Mycocentrospora acerina*) and/or sporulation in some species.
- Water pH: The pH of stream water can affect the formation of synanamorphs and hence sexual reproduction. Testing with different pH values including those of the source stream is thus recommended.
- Substrate/water volume ratio: this must be kept low in incubations of field material when water is not changed, in order to delay microbiologically induced staling; but also in pure cultures submerged in unchanged water, in order to delay chemically induced staling.

Methods used for anamorph induction other than the technique introduced in Sections 3.2 and 3.7 include: (1) Incubation of colonized leaves in continuously renewed distilled water (Bengtsson 1992); (2) incubation of pure cultures on 0.1% malt extract agar or even water agar; (3) Incubation of field material and pure cultures in periodically renewed nutrient solution (e.g. Ciferri 1959, Suberkropp 1991); (4) continuous water drip over pure cultures (Marvanová 1968); (5) continuous flow chambers (Descals et al. 1976, Descals & Moralejo 2001).

5. GLOSSARY

Anamorph	A supposedly asexual morph; some species have two or more anamorphs, which may appear simultaneously; cf. teleomorph, morph, synanamorph
Apothecium	Plural: apothecia. A type of teleomorphic fruit body characteristic of a major group of ascomycetes

Ascomycete	A major phylogenetic group of fungi
Ascospore	Meiospore of Ascomycetes
Basidiomycete	A major phylogenetic group of fungi
Basidiospore	Meiospore of Basidiomycetes
Conidium	Plural: conidia; cf. mitospore
Conidiophore	A part of a thallus bearing one or more conidia
Leptosphaeriaceous	Refers to the Ascomycete family Leptosphaeriaceae
Meiospore	A spore whose ontogeny involves at least a meiotic division; cf. mitospore
Microsclerotium	Plural: microsclerotia. Small sclerotium
Mitospore or conidium	A spore whose ontogeny is not known to involve meiosis; cf. meiospore
Morph or form	A part of a thallus which is usually associated with sexual or asexual reproduction
Mycelium	Plural: mycelia, a mass of filaments (called hyphae), the thallus of most fungi
Pseudothecium	Plural: pseudothecia. A type of teleomorphic fruit body characteristic of a major group of Ascomycetes
Sclerotium	Plural: sclerotia. A discrete, firm mass of hyphae or cells functioning as a resting body in certain fungi. It may give rise to a fruit body or mycelium.
Synanamorph	Any one of two or more anamorphs formed by some fungal species
Teleomorph	The sexual morph of a fungus

6. REFERENCES

Ando, K. & Tubaki, K. (1984). Some undescribed hyphomycetes in the rain drops from intact leaf-surface. *Transactions of the Mycological Society of Japan*, 25, 21-37.

Bärlocher, F. (ed.) (1992). *The Ecology of Aquatic Hyphomycetes*, Ecological Studies Vol. 94, Spirnger-Verlag, Berlin.

Bärlocher, F. & Kendrick, B. (1974). Dynamics of the fungal population on leaves in a stream. *Journal of Ecology*, 62, 761-791.

Bengtsson, G. (1992). Interactions between fungi, bacteria and beech leaves in a stream microcosm. *Oecologia*, 89, 542-549.

Chamier, A.-C. & Dixon P.A. (1982) Pectinases in leaf degradation by aquatic hyphomycetes I - The field-study. The colonization-pattern of aquatic hyphomycetes on leaf packs in a Surrey stream. *Oecologia*, 52, 109-115.

Chauvet, E. & Suberkropp, K. (1998) Temperature and sporulation of aquatic hyphomycetes. *Applied and Environmental Microbiology*, 64, 1522–1525.

Ciferri, R. (1959). Ecological observations on aquatic hyphomycetes. Omagiu lui Traian Savulescu cu prilejui implinirii a 70 de ani, Academia Republicii Populâre Romîne. Editura Academici Republicii Populâre Romine, pp.157-162.

Descals, E. (1987). Muestreo preliminar de hongos ingoldianos de Cataluña. *Revista Ibérica de Micología*, 4, 17-32.

Descals, E. (1997). Ingoldian fungi: some field and laboratory techniques. *Bolletí de la Societat d'Historia Natural de les Balears*, 40, 169-221.

Descals, E. (1998). Streamborne fungi from Karrantza (Basque Country) and surroundings. *Bolletí de la Societat d'Historia Natural de les Balears*, 41, 191-219.

Descals, E. & Moralejo, E. (2001). Water and asexual reproduction in the Ingoldian fungi. *Botanica Complutenses*, 25, 13-71.

Descals, E., Nawawi, A. & Webster, J. (1976). Developmental studies in *Actinospora* and three similar aquatic hyphomycetes. *Transactions of the British Mycological Society*, 67, 207-222.

Descals, E. & Rodríguez, J. (2002). *Cuadernos de Trabajo de Flora Micológica Ibérica 18. Bases corológicas*. Consejo Superior de Investigaciones Científicas. Madrid.

Fisher, P.J. & Petrini, O. (1989). Two aquatic hyphomycetes as endophytes in *Alnus glutinosa* roots. *Mycological Research*, 92, 367-368.

Gessner, M.O., Bärlocher, F. & Chauvet, E. (2003). Qualitative and quantitative analyses of aquatic hyphomycetes in streams. In: C.K.M. Tsui & K.D. Hyde, (eds.), *Freshwater Mycology*. Fungal Diversity Press, Hong Kong, pp. 127-157.

Ingold, C.T. (1942). Aquatic hyphomycetes of decaying alder leaves. *Transactions of the British Mycological Society*, 25, 339-417.

Ingold, C.T. & Chapman, B. 1952. Aquatic Ascomycetes: *Loramyces juncicola* Weston and *L. macrospora* n. sp. *Transactions of the British Mycological Society*, 35, 268-272.

Ingold, C.T. (1975). An illustrated guide to aquatic and waterborne hyphomycetes (Fungi Imperfecti) with notes on their biology. Freshwater Biological Association Scientific Publications 30, 1-96.

Iqbal, S.H. & Webster, J. (1973). Aquatic hyphomycete spora of the River Exe and its tributaries. *Transactions of the British Mycological Society*, 61, 331-346.

Kirby, J.J.H. (1987). A comparison of serial washing and surface sterilization. *Transactions of the British Mycological Society*, 88, 559-562.

Kirk, P.M., Cannon, P.F., David, J.C. & Stalpers, J.A. (2001). *Dictionary of the Fungi*. CAB International, Wallingford.

Marvanová, L. (1968). *Lemonniera centrosphaera* sp. nov. *Transactions of the British Mycological Society*, 51, 613-616.

Marvanová, L. (1997). Freshwater Hyphomycetes: A Survey with Remarks on Tropical Taxa. In K.K Janardhanan, C. Rajendran, K. Natarajan, & D.L. Hawksworth (eds.). Tropical Mycology. pp. 169-226. Science Publishers, Inc., Enfield NH.

Nawawi, A. (1985). Aquatic hyphomycetes and other water-borne fungi from Malaysia. Malayan Nature Journal 39, 75-134.

Shearer, C.A. & Webster, J. (1985). Aquatic hyphomycete communities in the River Teign. III. Comparison of sampling techniques. *Transactions of the British Mycological Society*, 84, 509-518.

Suberkropp, K. (1991). Relationships between growth and sporulation of aquatic hyphomycetes on decomposing leaf litter. *Mycological Research*, 95, 843-850.

Webster, J. & Descals, E. (1981). Morphology, distribution, and ecology of conidial fungi in freshwater habitats. In: G.T. Cole & B. Kendrick (eds.), *Biology of Conidial Fungi*, Vol 1. Academic Press, New York.

Willoughby, L.G. & Archer, J.F. (1973). The fungal spora of a freshwater stream and its colonization pattern on wood. *Freshwater Biology*, 3, 219-239.

CHAPTER 20

MAINTENANCE OF AQUATIC HYPHOMYCETE CULTURES

LUDMILA MARVANOVÁ

Masaryk University, Faculty of Science, Czech Collection of Microorganisms, Tvrdého 14, 616 00 Brno, Czech Republic.

1. INTRODUCTION

Pure cultures of fungi provide important information on morphological and physiological features necessary for reliable identification. Gross colony characters, such as texture, radial growth rate, colour of the front and back side, pigments in the agar medium, and the presence of sclerotia, can be observed by the unaided eye on solid agar media, whereas microscopic features are studied mostly in submerged cultures: microconidial synanamorphs, conidiophores, conidiogenous cells, the events of conidiogenesis and conidial morphogenesis, morphology of mature detached conidia, the presence and kind of chlamydospores and hyphopodia. All these characters may contribute to the accurate classification of a specimen (Chapter 21).

Biochemical and molecular identification methods such as monoclonal antibody-based immunoassays (e.g. Bermingham et al. 2001) and fluorescence *in situ* hybridization (McArthur et al. 2001) likewise require work with pure cultures, as do various kinds of ecological experiments (Bärlocher & Kendrick 1973, Arsuffi & Suberkropp 1983, Suberkropp 1991). Furthermore, molecular studies aimed at finding teleomorph-anamorph connections and phylogenetic relationships of mitosporic taxa or strain variation are also best based on pure cultures (Marvanová et al. 2002, Nikolcheva & Bärlocher 2002). Finally, several antibiotic substances produced by aquatic hyphomycetes were discovered in pure cultures (e.g. Gulis & Stephanovich 1999, Oh et al. 1999).

Pure cultures of taxa unknown to the investigator and of taxa difficult to identify on the basis of detached conidia should be established when starting an ecological study involving species identification in a previously unexplored area. It is recommended to keep voucher specimens and/or picture documentation of at least

M.A.S. Graça, F. Bärlocher & M.O. Gessner (eds.), Methods to Study Litter Decomposition: A Practical Guide, 143 – 152.

those species whose identification was ambiguous. Collaboration with a taxonomist is sometimes unavoidable, because even well studied areas contain many species that are still undescribed or difficult to identify. For example, Marvanová (2001) cited about 20% undescribed taxa in a protected area in the Czech Republic and Gönczöl et al. (2001) noted some 40% undescribed taxa in a single small stream in Hungary.

Ecologists may benefit from a reference species collection for comparative purposes. These must be kept alive at least for the duration of the study. If they prove important from some point of view for future work or are referred to in publications, they should be deposited in public culture collections of microorganisms (Smith & Onions 1994), to be generally available.

Culture collections have become especially important after the Convention on Biological Diversity was ratified in many states (Hawksworth 1996). They ensure conservation of microbial biodiversity outside the natural habitat (Glowka 1996). Microorganisms are understood as microbial genetic resources and should be protected on a national basis like other natural resources (Anonymous 1996). New or extremely rare microorganisms may not be restricted to those pristine locations that are usually favoured for nature conservation.

1.1. Location of Culture Collections and Information about Strains

General information on culture collections and their holdings can be obtained from the World Data Centre for Microorganisms, a component of the World Federation of Culture Collections home page (http//www.wdcm.nig.ac.jp). Strains of aquatic hyphomycetes are currently preserved in the following major culture collections: Centraalbureau voor Schimmelcultures (http//www.cbs.knaw.nl), CABI Bioscience Genetic Resource Collection (http//www.cabi-bioscience.org/htmlgrc.htm), BCCM™/MUCL (Agro)Industrial Fungi & Yeasts Collection (http//www.belspo.be/bccm/mucl/htm), and American Type Culture Collection (http//www.atcc.org).

Maintenance of a culture means keeping it viable, pure, authentic and genetically unchanged. Long-term maintenance implies preserving cultures for one to tens of years without reinoculation. This can be achieved by reducing or even suspending fungal metabolism. There are many conservation methods to effect this, but none is universally applicable for all microorganisms (Kirsop & Doyle 1984, Stalpers et al. 1987, Smith & Onions 1994, Kolkowski & Smith 1995, Hubálek 1996, Smith 1998). As a rule, aquatic hyphomycetes do not produce aerial macroconidia on solid media. Therefore, methods allowing conservation of nonsporulating mycelia should be employed. For all forms of conservation it is necessary to start with well grown, vigorous cultures, preferably at the early to mid stationary phase of growth. They should be cultivated at optimum conditions for growth.

1.2. Conservation at Reduced Metabolism in Deionized Water

This method was first used by Castellani (1939, 1967) for human pathogenic fungi, successfully adopted for various groups of fungi (e.g. Ellis 1979) and is still in use in the CABI Bioscience Genetic Resource Collection (Smith & Onions 1994). Reduced fungal metabolism during storage is probably caused by low nutrient concentrations in the medium and by restricted oxygen availability under conditions of submergence.

Advantages: Cheap equipment, cultures require little space for storage, morphology and sporulation capacity are usually well preserved, transfer to fresh agar media is generally sufficient to revive culture.

Disadvantages: Danger of genome change and/or contamination during storage. The shelf life is generally 2—5 years, although decreased viability after 20 years of storage was encountered with some isolates of aquatic hyphomycetes in the Czech Collection of Microorganisms (unpubl. data).

1.3. Conservation at Reduced Metabolism under Mineral (Paraffin) Oil

The first report on extensive use of this method is by Buell & Weston (1947) and further successful applications were reported by Fennell (1960) and Onions (1971). The method is still in use in some large culture collections.

Advantages: Cheap equipment, cultures require little space for storage, the sporulation capacity is moderately well preserved. The shelf life is usually 5—10 years, occasionally even up to 30 years.

Disadvantages: Reviving must be done by repeated transfer because of the presence of oil on the inoculum, danger of genome change and/or contamination during storage.

1.4. Conservation at Suspended Metabolism

Cryopreservation is achieved by storage in polypropylene straws in liquid nitrogen (LN) vapours at −130 to −170 °C. This is based on the classical method of long-term maintenance of fungi in glass vials in LN developed in the 1970s (Elliott 1976), which has been successfully adopted by several large culture collections (e.g. Stalpers et al. 1987, Hoffman 1992) for a wide range of both filamentous fungi and yeasts.

Advantages: Minimum genome and phenotype changes, excellent preservation of sporulation capacity, requires little space, long shelf life (tens of years) and contamination during storage practically excluded.

Disadvantages: Relatively expensive equipment and rather high maintenance costs, comparatively laborious procedure, requires regular and reliable LN supply, potential health hazards during manipulation with LN.

2. EQUIPMENT, CHEMICALS AND CULTIVATION MEDIUM

2.1. Equipment and Material

- Glass bottles with screw caps, approx. 10 ml capacity
- Glass test tubes with inner diameter of ≥12 mm, and cotton plugs, or pressure resistant plastic screw-cap bottles of at least 30 ml capacity, preferably wide-necked
- Adhesive labels, permanent markers, pens resistant to LN
- Inoculation needles or hooks
- Pasteur pipettes
- Burners
- Polypropylene or polyvinyl chloride drinking straws, inner diameter 2—3 mm, pressure resistant
- Forceps
- Scalpels
- Cork borers
- Polypropylene cryotubes, ca. 2 ml capacity, pressure resistant (e.g. Intermed NUNC, Denmark)
- Canes for holding a set of cryotubes
- Cylindrical holders for canes
- Storage container for LN
- Petri dishes, plastic or glass
- Biohazard II laminar flow cabinet
- Autoclave
- Laboratory Drying Oven

2.2. Chemicals

- Deionized water
- Mineral oil (medicinal paraffin oil, density of 0.83—0.89 g cm^{-3})
- Glycerol, analytical grade
- Liquid nitrogen

2.3. Cultivation Medium

- Malt agar 2% (e.g. Difco, Oxoid or other)

3. PROCEDURES

3.1. Preservation in Deionized Water

1. Prepare Petri dishes with vigorous, well grown pure cultures.
2. Sterilize bottles with deionized water (ca. 5 ml in each) with caps loosely screwed on. Cool to room temperature.
3. With a sterilized cork borer (diameter approx. 8 mm), or the broad end of a Pasteur pipette, aseptically cut disks from a colony. Alternatively, cut small squares with a sterile scalpel.
4. Transfer several disks or squares to a bottle with sterile water and tightly screw on cap.
5. The basic unit in a culture collection is a strain (= isolate). Label each bottle with strain number (numbers are preferable to species names) and date.
6. Store at 10—18 °C in dark.
7. Revive culture by aseptically lifting a disk or square of the culture from the bottle and placing it mycelium down on a fresh agar medium plate.

3.2. Preservation under Paraffin Oil

1. Prepare sterile glass test tubes with cotton plugs or bottles with screw caps containing slant agar medium: pour agar medium into tubes or bottles and keep them in tilted position until the medium solidifies to get the slant surface.
2. Inoculate the agar with a culture. Use several pieces of inoculum and distribute them over the entire slant surface when handling slow-growing cultures with restricted colonies.
3. Incubate until the agar surface is well covered with mycelium (usually 1—2 weeks at 15—20 °C).
4. Prepare test tubes with appropriate amount of paraffin oil (it must go well above the top of the slant agar medium), one for each culture.
5. Autoclave paraffin oil in test tubes (121 °C for 20 min) and keep them at 160 °C in laboratory dryin oven for 2 h. Alternatively, autoclave twice (121 °C for 15 min).
6. Pour the cooled paraffin oil aseptically into tubes or bottles with cultures, always using one tube with paraffin oil per tube or bottle with a culture. The paraffin oil meniscus should be ca. 1 cm above the highest point of the agar slope or fungal growth.
7. Label each tube or bottle with strain number and date.
8. Store at 10 — 18 °C in the dark.
9. For retrieval, aseptically cut out a piece of the culture with an inoculation needle or hook, drain the paraffin oil off by pressing the piece against the tube wall and place it on fresh agar medium. Incubate at 15—20 °C.

10. Keep the inoculated Petri dishes tilted at a 45° angle. This allows the excess paraffin oil to flow down. The first subculture is usually slimy, and a second transfer is often necessary.

3.3. Preservation in Liquid Nitrogen (LN) Vapours (Figs 20.1 and 20.2)

1. Prepare cultures in Petri dishes with agar medium that was supplemented with 5% (w/v) of glycerol before autoclaving. Alternatively, flood the culture with 10% sterile glycerol (w/v in distilled water) 1 h before processing.
2. Cut drinking straws into 25-mm pieces and sterilize in autoclave (121 °C, 15 min) or with gamma-radiation.
3. Hold one straw piece with sterile forceps and punch out a disk from the agar culture. Repeat the procedure until the straw is filled.
4. Place 4 or 5 straws filled with one strain into one cryotube and label it with the strain number and date. To differentiate easily between strains, various colours of caps and straws may be used.
5. Place the cryotubes into canes. Store canes with cryotubes at –70 °C for 2 h, protected by insulation in polystyrene boxes. Then place the canes into cylindrical holders and these into the LN vapour phase. Wear full face, hand and clothing protection during all work with LN. There is also a danger of anoxia resulting from possible oxygen deficiency caused by vaporization of LN.
6. Holders and canes must be numbered. For easy retrieval, prepare a diagram showing the location of cryotubes with straws within the canes and of canes within the holders.
7. For revival of a strain, remove a cryotube from the cane and put it into a polystyrene block in a box. This will prevent thawing for ca. 15 min. Open the cryotube and place one straw aseptically on a Petri dish with 2% malt agar. Incubate at 15—20 °C until growth appears from the open ends of the straw.
8. Transfer the growing culture to a fresh slant agar. Alternatively, place the straw directly on a slant agar and incubate as above. However, aquatic hyphomycetes usually recover better when first placed on Petri dishes.

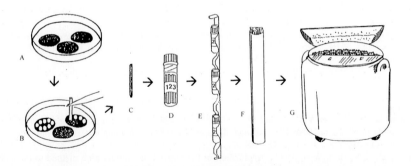

Figure 20.1. Cryoperservation of fungi. A – pure culture of filamentous fungi; B –punching the culture with a straw; C – straw filled with pieces of culture; D – cryotube with straws; E – cryotubes clamped to aluminium cane; F – cylindrical holder for canes; G – liquid nitrogen container.

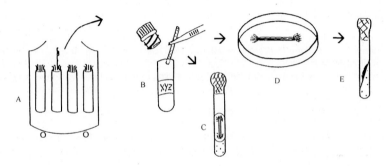

Figure 20.2. Removal and thawing of frozen cultures. A – liquid nitrogen container; B – removing one straw from the cryotube; C – incubating the straw directly on slant agar; D – incubating the straw on agar medium in a Petry dish; E – Subculture from the straw on slant agar.

4. REFERENCES

Anonymous (1996). *Information Document on Access to Ex-situ Microbial Genetic Resources within the Framework of the Convention on Biological Diversity.* World Federation for Culture Collections, Yata, Mishima, Shizuoka, Japan, 25 pp.

Arsuffi, T.L. & Suberkropp, K. (1985). Selective feeding by stream caddisfly (Trichoptera) detritivores on leaves with fungal-colonized patches. *Oikos*, 45, 50–58.

Bärlocher, F. & Kendrick, B. (1973). Fungi in the diet of *Gammarus pseudolimnaeus* (Amphipoda). *Oikos*, 24, 295–300.

Bermingham, S., Dewey, F.M., Fisher, P.J. & Maltby, L. (2001). Use of a monoclonal antibody based immunoassay for the detection and quantification of *Heliscus lugdunensis* colonizing alder leaves and roots. *Microbial Ecology*, 42, 506–512.

Buell, C.B. & Weston, W.H (1947). Application of the mineral oil conservation method to maintaining collections of fungus cultures. *American Journal of Botany*, 34, 555–561.

Castellani, A. (1939). Maintenance and cultivation of common pathogenic fungi of man in sterile distilled water. *Journal of Tropical Medicine and Hygiene*, 42, 225–226.

Castellani, A. (1967). Maintenance and cultivation of common pathogenic fungi of man in sterile distilled water. Further researches. *Journal of Tropical Medicine and Hygiene*, 70, 181–184.

Elliott, T.J. (1976). Alternative ampoule for storing fungal cultures in liquid nitrogen. *Transactions of the British Mycological Society*, 67, 545-546.

Ellis, J.J. (1979). Preserving fungus strains in sterile water. *Mycologia*, 71, 1072–1075.

Fennell, D.I. (1960). Conservation of fungus cultures. *Botanical Review*, 26, 79–141.

Glowka, L. (1996). The Convention on Biological Diversity: issues of interest to the microbial scientist and microbial culture collections. In: R.A. Samson, J.A. Stalpers, D. van der Mei, & A.H. Stouthamer (eds.), *Culture Collections to Improve the Quality of Life* (pp. 36–60). Centraalbureau voor Schimmelcultures. Baarn.

Gönczöl, J., Révay, Á. & Csontos, P. (2001). Effect of sample size on the detection of species and conidial numbers of aquatic hyphomycetes collected by membrane filtration. *Archiv für Hydrobiologie,* 150, 677–691.

Gulis, V.I. & Stephanovich, A.I. (1999). Antibiotic effects of some aquatic hyphomycetes. *Mycological Research*, 103, 111–115.

Hawksworth, D.L. (1996). Microbial collections as a tool in biodiversity and biosystematic research. In: R.A. Samson, J.A. Stalpers, D. van der Mei, & A.H. Stouthamer (eds.), *Culture Collections to Improve the Quality of Life* (pp. 26–35). Centraalbureau voor Schimmelcultures. Baarn.

Hoffman, P. (1992). Technical Information Sheet No. 5. Cryopreservation of fungi. In: K. A Malik (ed.), *Technical Information for Culture Collections. Curators in Developing Countries* (pp. 17–20). DSM-Deutsche Sammlung für Mikroorganismen und Zellkulturen. Braunschweig.

Hubálek, Z. (1996). *Cryopreservation of Microorganisms at Ultra-low Temperatures.* Academia, Prague.

Kirsop, B.A. & Doyle, A. (eds.). (1991). *Maintenance of Microorganisms and Cultured Cells. A Manual of Laboratory Methods*, 2nd ed. Academic Press. London.

Kolkowski, J.A., & Smith, D. (1995). Cryopreservation and freeze-drying of fungi. In: J. G. Day, & M. R. McLellan (eds.), *Cryopreservation and Freeze-Drying Protocols.* (pp. 49–61), Humana Press. Totowa.

Marvanová, L. (2001). Streamborne fungal spora in running waters of the Bohemian Forest. *Silva Gabreta,* 7, 147–164.

Marvanová, L., Landvik, S., Fisher, P.J., Moss, S.T. & Ainsworth, A.M. (2002). A new fungus with arthroconidia from foam. *Nova Hedwigia*, 75, 255–269.

McArthur, F.A., Baerlocher, M.O., MacLean, N.A.B., Hiltz, M.D. & Bärlocher, F. (2001). Asking probing questions: can fluorescent in situ hybridization identify and localise aquatic hyphomycetes on leaf litter. *International Review of Hydrobiology*, 86, 429–438.

Nikolcheva, L.G. & Bärlocher, F. (2002). Phylogeny of *Tetracladium* based on 18S rDNA. *Czech Mycology*, 53, 285–295.

Oh, H., Kwon, T.O., Gloer, J.B., Marvanová, L. & Shearer, C.A. (1999). Tenellic acids A-D: new bioactive diphenyl ether derivatives from the aquatic fungus *Dendrospora tenella*. *Journal of Natural Products*, 62, 580–583.

Onions, A.H.S. (1971). Preservation of fungi. In: C. Booth (ed.), *Methods in Microbiology* (pp. 113–151). Academic Press. London.

Smith, D. (1998). The use of cryopreservation in the ex-situ conservation of fungi. *Cryo-Letters*, 19, 79–90.

Smith, D. & Onions, A.H.S. (1994). *The Preservation and Maintenance of Living Fungi*. 2nd ed. CAB International. Wallingford.

Stalpers, J.A., de Hoog, S. & Vlug, I.J. (1987). Improvement of the straw technique for the preservation of fungi in liquid nitrogen. *Mycologia*, 79, 82–89.

Suberkropp, K. (1991). Relationships between growth and sporulation of aquatic hyphomycetes on decomposing leaf litter. *Mycological Research*, 95, 843–850.

CHAPTER 21

AN ILLUSTRATED KEY TO THE COMMON TEMPERATE SPECIES OF AQUATIC HYPHOMYCETES

VLADISLAV GULIS[1], LUDMILA MARVANOVÁ[2] & ENRIQUE DESCALS[3]

[1]*Vladislav Gulis, Instituto do Mar, Universidade de Coimbra, Coimbra 3004-517, Portugal; Current address: Department of Biological Sciences, Box 870206, University of Alabama, Tuscaloosa, AL 35487, USA;* [2]*Ludmila Marvanová, Czech Collection of Microorganisms, Faculty of Science, Masaryk University, Tvrdého 14, 602 00 Brno, Czech Republic;* [3]*Enrique Descals, IMEDEA, c. Miquel Marquès 21, 07190 Esporles, Mallorca, Spain.*

1. INTRODUCTION

Aquatic hyphomycetes play a key role in the decomposition of allochthonous plant litter and food webs in lotic ecosystems (e.g. Suberkropp & Klug 1976, Bärlocher 1992, Hieber & Gessner 2002). Soon after colonizing a substrate, many species produce vast amounts of conidia that enter the water column and are transported downstream. Aquatic hyphomycetes can invest up to 80% of their production into sporulation and conidial production alone has been shown to account for up to 8—12% of leaf litter mass loss (Suberkropp 1991). Most aquatic hyphomycetes form tetraradiate, variously branched or scolecoid (worm-like) conidia that are adapted for dispersal in flowing water (Webster & Descals 1981). Since conidia are mostly characteristically shaped, it is often possible to identify them to species, count them and thus gain insight into the structure of the fungal community developing on submerged substrates (Bärlocher 2004). This facilitates ecological studies that link fungal biodiversity with functional aspects of ecosystems such as organic matter decomposition.

The objective of this chapter is to provide assistance for the fast identification of aquatic hyphomycete conidia in ecological studies carried out in temperate climates. Conidia can be sampled in transport (water, foam), from naturally colonized

M.A.S. Graça, F. Bärlocher & M.O. Gessner (eds.), Methods to Study Litter Decomposition:
A Practical Guide, 153 – 168.

submerged substrates after inducing sporulation in the laboratory or from pure cultures (Gessner et al. 2003; Chapters 19 and 24).

It is important to acknowledge that the presented key includes only 64 of ca. 170 species of aquatic hyphomycetes occurring in temperate climate (over 300 species described worldwide). Along with common species we also included some less frequent ones, whose conidia may be confused with others. The key is primarily based on morphological characters of detached conidia; however, for species with similar or less differentiated conidia, their mode of conidiogenesis, which may also be diagnostic, is illustrated. Only typical conidia developed under submerged or semi-submerged conditions are considered. Even though we include drawings of conidia for all treated species, we encourage the reader to consult with taxonomic specialists and additional pertinent literature (e.g. Petersen 1962, 1963a,b, Nilsson 1964, Dudka 1974, Ingold 1975, Webster & Descals 1981, Marvanová 1997) as well as original species descriptions since some aquatic hyphomycetes (especially scolecosporous species) cannot be identified with certainty on the basis of detached conidia only. A glossary of some terms that may cause difficulties follows the key.

2. KEY TO THE COMMON TEMPERATE SPECIES OF AQUATIC HYPHOMYCETES (BASED ON CONIDIA)

1. Conidia variously branched or appearing tri-, tetra- or multiradiate.................2

1a. Conidia of simple shape (scolecoid, globose, ellipsoid, fusiform, clavate, etc., some with short outgrowths or basal extensions) 50

2. Conidia appearing triradiate..3

2a. Conidia of different morphology 4

3. Conidia spanning 8—13 µm, 3-celled, ends obtuse .. *Tricellula aquatica* (Fig.1)

3a. Conidial span 20—46 µm, apices acute ....*Ypsilina graminea* (Fig.2)

4. Conidia small, spanning up to 25 µm, outline triangular, with a short axis and 3 laterals...*Lateriramulosa uniinflata* (Fig. 3)

4a. Conidia of different shape...5

5. Conidia with a clamp connection on axis 6

5a. Clamp connections absent 7

6. Conidial elements cylindrical, axis gently curved or sigmoid, with an excentric basal extension, branch insertions unconstricted ....
.... *Taeniospora gracilis* var. *enecta* (Fig. 4a)

6a. Conidial elements long-fusoid, axis strongly curved or sigmoid, basal extension absent, branch insertions subconstricted...
.... *Taeniospora gracilis* var. *gracilis* (Fig. 4b)

7. Conidia relatively large with typically numerous primary and secondary (sometimes tertiary) branches, elements cylindrical...8

7a. Conidia of different shape..11

8. Conidia resembling a fir tree with a more or less straight axis and perpendicular branches tending to aggregate near its base 9

8a. Conidial elements gently curved, branches distributed along the axis length, caducous (breaking off readily) 10

9. Conidia ca. 200 μm long, with more than 15 branches, elements 4—5 μm wide ....*Dendrospora erecta* (Fig. 5)

9a. Conidia with up to 14 branches, elements 3—4 μm wide..................................... ....*Dendrospora tenella* (Fig. 6)

10. Branches typically on one side of the axis, branch insertions abruptly constricted..*Varicosporium elodeae* (Fig. 7)

10a. Conidial elements delicate, gently constricted at septa, branches on both sides of the axis, insertions gradually narrowed ..... *Varicosporium delicatum* (Fig. 8)

11. Conidia with an axis, 1—2 primary and one secondary branch, elements tapering distally 12

11a. Conidia of different shape..15

12. Conidial elements straight or nearly so, branching dorsal 13

12a. Conidial elements strongly curved or sigmoid, branching ventral...................14

13. Axis 26—51 μm long, 3—4(—5) septa ...... *Pleuropedium tricladioides* (Fig. 9)

13a. Axis 38—75(—100) μm long, elements multiseptate ....*Pleuropedium multiseptatum* (Fig. 10)

14. Conidia with one primary and one secondary branch... ....*Gyoerffyella gemellipara* (Fig. 11)

14a. Conidia with two primary and one secondary branch.. .... *Gyoerffyella rotula* (Fig. 12)

15. Conidia with recurved axis and 2—3 branches (4—5 ends) 16

15a. Conidia of different shape..17

16. Conidia with up to 5 ends, stout .... *Tripospermum myrti* (Fig. 13)

16a. Conidia with up to 4 ends, more slender in appearance....................................... ....*Tripospermum camelopardus* (Fig. 14)

17. Conidia spanning over 70 μm, with 4 long filiform extensions, conidial body of 2 parts ....*Campylospora chaetocladia* (Fig. 15)

17a. Conidia of different shape..18

18. Conidia with a stalk bearing elements of 2 different shapes: filiform (subulate) and digitiform, filiform (subulate) and globose or only digitiform...................19

18a. Conidia of different morphology 23

19. Conidia with 2 globose and 3 filiform (subulate) elements ....
.... *Tetracladium marchalianum* (Fig. 16)

19a. Conidia with filiform (subulate) and digitiform or only digitiform elements .. 20

20. Filiform (subulate) elements lacking, 2 digitiform elements furcate, conidia with 6 apices .... *Tetracladium apiense* (Fig. 17)

20a. Both filiform (subulate) and digitiform elements present 21

21. Conidia with 2(—3) filiform (subulate) and 2 digitiform elements, basal extension absent .... *Tetracladium maxilliforme* (Fig. 18)

21a. Conidia with 5—6 apices, basal extension typically present 22

22. Conidia with 2 digitiform and 3 filiform (subulate) elements ....
.... *Tetracladium furcatum* (Fig. 19)

22a. Conidia with 3 digitiform and 3 filiform (subulate) elements ....
.... *Tetracladium setigerum* (Fig. 20)

23. Conidia with relatively broad body (clavate, fusoid, etc.) and 2—4 branches or if all elements are of similar width then conidia spanning up to 15 μm 24

23a. Conidia with all elements of similar width, spanning more than 25 μm 30

24. Conidial body clavate, with 3 coronate branches 25

24a. Conidial body long-clavate, navicular or obclavate with one terminal and 2—3 lateral branches 28

25. Conidial body 10—15 μm long, with 3 conoid branches, conidia appear stellate
.... *Heliscella stellata* (Fig. 21)

25a. Conidia larger, branches filiform 26

26. Conidial body 35—50 × 10 —12 μm .... *Clavariopsis aquatica* (Fig. 22)

26a. Conidial body 3—5 μm wide 27

27. Conidial body 15—25 μm long .... *Clavatospora longibrachiata* (Fig. 23)

27a. Conidial body 45—70 μm long .... *Heliscus tentaculus* (Fig. 24)

28. Conidial body navicular to obclavate, straight, lateral branches 2—3, basal extension absent .... *Naiadella fluitans* (Fig. 25)

28a. Conidial body long-clavate, curved distally, lateral branches 2, basal extension percurrent, conidia have "mosquito" or "penguin" appearance 29

29. Conidial body hyaline, branches 40—80(—100) μm long ....
.... *Culicidospora aquatica* (Fig. 26)

29a. Conidial body hyaline or subfuscous, branches less than 40 μm long ....
.... *Culicidospora gravida* (Fig. 27)

30.   Conidia tetraradiate in a broad sense, i.e. appearing as 4 arms radiating from a common point or from a central cell, or basiverticillate, or with terminal branches on a stalk, or with paired or subopposite laterals on a geniculate or curved axis..31

30a. Conidia with elongate axis and 2 alternate, not subopposite, branches...........45

31.   Conidia with a distinct globose central cell and 4 radiating arms....................32

31a. Conidia of different shape...33

32.   Arms as broad as central cell, long-obclavate, insertions constricted.................
      ....*Lemonniera pseudofloscula* (Fig. 28)

32a. Arms thinner than central cell, cylindrical with subclavate ends, insertions unconstricted...*Lemonniera centrosphaera* (Fig. 29)

33.   Conidia with 4 arms radiating from a common point, i.e. truly tetraradiate (indistinct central cell sometimes present)..34

33a. Conidia of different morphology 35

34.   Arms cylindrical, 50—100 × 3—4 μm................*Lemonniera aquatica* (Fig. 30)
34a. Arms conoid or obclavate, 20—45 × 4—9 μm ..*Lemonniera terrestris* (Fig. 31)

35.   Conidia basiverticillate 36

35a. Conidia with a stalk bearing terminal branches or with paired or subopposite laterals on a geniculate or curved axis 38

36.   Conidial axis up to 105 μm long, elements cylindrical, branch insertions constricted.. *Lemonniera filiformis* (Fig. 32)

36a. Conidial axis long-obclavate, up to 70 μm long...37

37.   Axis 2-celled, lower cell often inflated, branches cylindrical, insertions subconstricted, septa indistinct or lacking ....
      .... *Triscelophorus monosporus* (Fig. 33)

37a. Axis and branches long-obclavate, multiseptate, branch insertions abruptly constricted... *Triscelophorus acuminatus* (Fig. 34)

38.   Conidia with 3 terminal branches (or 2, one of them forking again), elements constricted at insertions .... *Articulospora tetracladia* (Fig. 35)

38a. Conidia with geniculate or curved axis and 2 branches attached near its middle
     39

39.   Branches subopposite, axis subconstricted at a septum between branch insertions...40

39a. Branches paired, axis not constricted..41

40.   Axis typically over 90 μm long, elements cylindrical ....
      ....*Fontanospora eccentrica* (Fig. 36)

40a. Axis less than 90 µm long, branches long-obclavate..
....*Fontanospora fusiramosa* (Fig. 37)

41. Conidia spanning over 90 µm..42

41a. Conidia spanning less than 70 µm 43

42. Branches submedian, insertions subconstricted, axis slightly swollen and bent at branch insertions...*Geniculospora inflata* (Fig. 38)

42a. Elements of equal length, branches gently curved backwards, insertions unconstricted..*Tetrachaetum elegans* (Fig. 39)

43. Conidial elements straight, axis bent at branch insertions, lower part of axis often subclavate, distal part thinner, cylindrical, branch-like, often twice as long...*Stenocladiella neglecta* (Fig. 40)

43a. Conidial elements gently curved...44

44. Lower element of axis cylindrical to subclavate, distal elements narrow-obclavate, branch insertions strongly constricted ....
....*Alatospora pulchella* (Fig. 41)

44a. Elements cylindrical or long-fusoid or branches (0—2) subulate, insertions unconstricted to constricted ....*Alatospora acuminata* (Fig. 42)

45. Branch insertions unconstricted or subconstricted...46

45a. Branch insertions abruptly constricted..48

46. Axis 150—200 µm long, elements gently curved, branch insertions subconstricted ....*Tricladium chaetocladium* (Fig. 43)

46a. Axis up to 120 µm long 47

47. Axis 50—120 µm long, geniculate, branch insertions unconstricted.................
....*Tricladium angulatum* (Fig. 44)

47a. Axis 40—60 µm long, often curved in lower part, base swollen, branch insertions subconstricted................................*Tricladium curvisporum* (Fig. 45)

48. Axis geniculate or curved, elements cylindrical ... *Tricladium patulum* (Fig. 46)

48a. Axis fusoid or long-fusoid, straight or gently curved, branches long-obclavate..
....49

49. Axis 50—75 × 2.5—3.5 µm, apices acute........ *Tricladium attenuatum* (Fig. 47)

49a. Axis 60—140 × 5—7 µm....................................*Tricladium splendens* (Fig. 48)

50. Conidia scolecoid or filiform, i.e. length to width ratio >10..........................51

50a. Conidia of different shape...58

51. Detachment scar lateral, dorsal (or basal extension integrated), conidia aseptate, lunate or sigmoid.................................. *Lunulospora curvula* (Fig. 49)

51a. Detachment scar at the base of conidia, basal extension excentric, percurrent or absent, conidia septate, variously curved..52

52. Basal extension excentric...53

52a. Basal extension percurrent or lacking...55

53. Conidia filiform, 2.5—3.5 μm wide................*Anguillospora filiformis* (Fig. 50)

53a. Conidia 6—15 μm wide..54

54. Conidia hyaline, 110—190 × 6—13 μm, with 4—6 septa, one middle cell typically larger...*Mycofalcella calcarata* (Fig. 51)

54a. Conidia (or central cells) sometimes fuscous, 150—200 × 8—15 μm, with 7—11 septa, cells in the broad part of conidium of similar size..............................
....*Mycocentrospora acerina* (Fig. 52)

55. Conidia arcuate or sigmoid, 150—250 × 5—6 μm, 7—13 septate, basal extension growing through a frill (remnants of separating cell), which is usually difficult to observe ....*Anguillospora longissima* (Fig. 53)

55a. Frill at the base of conidia absent 56

56. Conidia sigmoid, long-fusoid, over 5 μm wide 57

56a. Conidia filiform, 90—120 × 1.5—2.5 μm........ *Flagellospora curvula* (Fig. 54)

57. Conidia 150—300 × 5—7 (—9) μm, 10—23 septate, base truncate or with a subulate extension..*Anguillospora furtiva* (Fig. 55)

57a. Conidia 120—180 × 8—14 μm, base truncate or with a blunt extension............
....*Anguillospora crassa* (Fig. 56)

58. Conidia isodiametric, clavate or fusiform to rhomboid, with short outgrowths
....59

58a. Conidia ellipsoid to reniform, without outgrowths, aseptate, (13—16—20(—24) × 8—10 μm ....*Dimorphospora foliicola* (Fig. 57)

59. Conidia globose, cubic to almost stellate in appearance, with 4—6 more or less equidistant, sometimes indistinct outgrowths 60

59a. Conidia clavate or fusiform to long-rhomboid 61

60. Conidia (10—)11—17(—21) μm in diam. ....*Goniopila monticola* (Fig. 58)

60a. Conidia (8—)9—13(—14) μm in diam. (conidia can also be limoniform to fusiform with 0—2 septa)............................. *Margaritispora aquatica* (Fig. 59)

61. Conidia fusiform to long-rhomboid, 3-celled, central cell inflated, with short equatorial outgrowths .... *Tumularia aquatica* (Fig. 60)

61a. Conidia clavate 62

62. Conidia curved, subclavate, 25—50 × 3—4 μm, outgrowths (1—)2 ....
.... *Heliscus submersus* (Fig. 61)

62a.   Conidia straight, outgrowths 0—4 63

63.   Distal cell swollen, outgrowths scattered over its surface, conidial body 25—40
       × 10—12 μm .... *Tumularia tuberculata* (Fig. 62)

63a.   Distal cell not inflated, outgrowths coronate 64

64.   Conidia broadly clavate to obcampanulate, 13—25 × 7—13 μm, conidial base
       with a denticle .... *Heliscina campanulata* (Fig. 63)

64a.   Conidia subclavate or clove-shaped, 20—45 × 4—6 μm, with 3(—4) conoid
       outgrowths or apex oblique .... *Heliscus lugdunensis* (Fig. 64)

3. GLOSSARY

arcuate   curved like a bow or arch.

basiverticillate   having similar elements arranged in a whorl at the lowermost
                       portion of parent element.

clavate   gradually broadening towards the distal part, club-shaped.

constricted   strongly and often abruptly narrowed.

coronate   having elements arranged in a crown-like fashion.

denticle   small tooth-like projection.

digitiform   finger-shaped.

excentric   located off the center (here refers to basal extensions on the side of
                       conidial scar).

filiform   resembling a thread or filament.

furcate   divided into two elements, forked.

fuscous   having a brownish gray color.

fusiform, fusoid   tapering towards each end, spindle-shaped.

geniculate   bent abruptly at an angle like a knee joint.

navicular   resembling a boat.

obclavate   gradually broadening towards the proximal part, *cf.* clavate.

obcampanulate   shaped like an inverted bell.

percurrent   here refers to basal extensions growing through a scar, *cf.*
                       excentric.

recurved   curved backwards.

reniform   kidney-shaped.

scar (conidial)   part of a septum involved in secession; it forms the base of
                       conidium, but sometimes it is replaced by percurrent basal
                       extension.

scolecoid   worm-like.

sigmoid   curved like the Greek letter sigma when standing at the end of a
                       word (ς) or the Latin letter S.

stellate   star-shaped, consisting of short elements radiating from a common
                       center.

sub	prefix signifying inferior position or degree: under, below, almost, not completely; *e.g. submedian* – situated below the middle, *subopposite* – arranged in pairs but not exactly on the same level.
subulate	tapering gradually to a point, awl-shaped.
tetraradiate	having four radiating elements.
triradiate	having three radiating elements.
truncate	terminating abruptly as if having the end cut off.

4. NOTES

Most of the drawings of conidia used in the key are by the authors (published or unpublished). Twenty-seven were taken from Bärlocher and Marvanová (2005). Illustrations of conidia from the genera *Dendrospora, Gyoerffyella* and *Tetracladium* are from Descals and Webster (1980), Marvanová (1975), and Roldán *et al.* (1989), respectively. *Anguillospora furtiva* is from the original description by Descals *et al.* (1998).

In order to identify species with conidia of simple shapes (e.g. *Dimorphospora foliicola, Goniopila monticola, Margaritispora aquatica,* many species with filiform conidia) details of conidiogenesis should be observed.

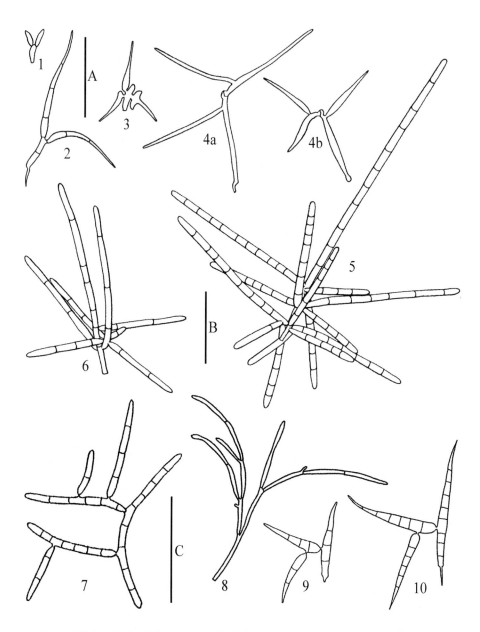

Figures 21.1—10. Conidia of aquatic hyphomycetes. 1. Tricellula aquatica. 2. Ypsilina graminea. 3. Lateriramulosa uniinflata. 4a. Taeniospora gracilis var. enecta. 4b. Taeniospora gracilis var. gracilis. 5. Dendrospora erecta. 6. Dendrospora tenella. 7. Varicosporium elodeae. 8. Varicosporium delicatum. 9. Pleuropedium tricladioides. 10. Pleuropedium multiseptatum. Scale bar A (Figs. 1—4) = 25 μm, B (Figs. 5—6) = 50 μm, C (Figs. 7—10) = 50 μm.

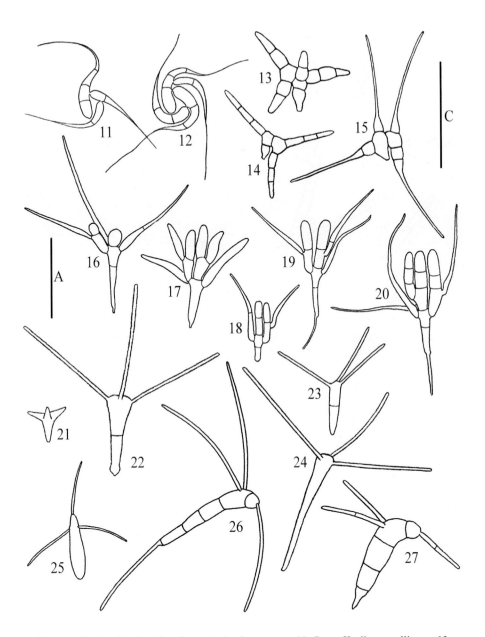

Figures 21.11—27. Conidia of aquatic hyphomycetes. 11. Gyoerffyella gemellipara. *12.*
Gyoerffyella rotula. *13.* Tripospermum myrti. *14.* Tripospermum camelopardus. *15.*
Campylospora chaetocladia. *16.* Tetracladium marchalianum. *17.* Tetracladium apiense. *18.*
Tetracladium maxilliforme. *19.* Tetracladium furcatum. *20.* Tetracladium setigerum. *21.*
Heliscella stellata. *22.* Clavariopsis aquatica. *23.* Clavatospora longibrachiata. *24.* Heliscus
tentaculus. *25.* Naiadella fluitans. *26.* Culicidospora aquatica. *27.* Culicidospora gravida.
Scale bar A (Figs. 11, 12, 16—21, 23) = 25 μm, C (Figs. 13—15, 22, 24—27) = 50 μm.

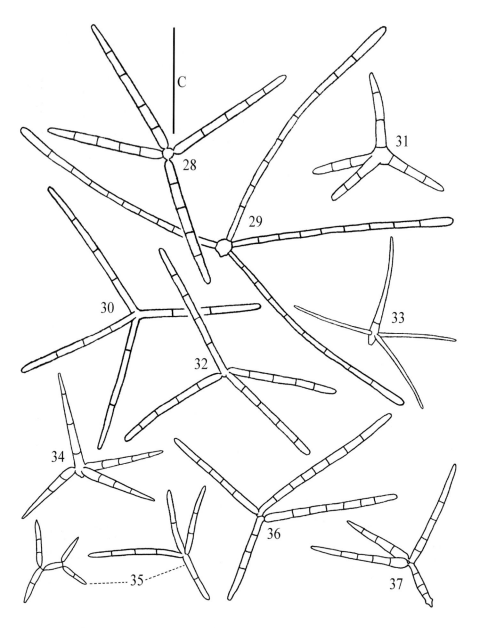

Figures 21.28—37. Conidia of aquatic hyphomycetes. 28. Lemonniera pseudofloscula. *29.* Lemonniera centrosphaera. *30.* Lemonniera aquatica. *31.* Lemonniera terrestris. *32.* Lemonniera filiformis. *33.* Triscelophorus monosporus. *34.* Triscelophorus acuminatus. *35.* Articulospora tetracladia. *36.* Fontanospora eccentrica. *37.* Fontanospora fusiramosa. *Scale bar C (Figs. 28—37) = 50 μm.*

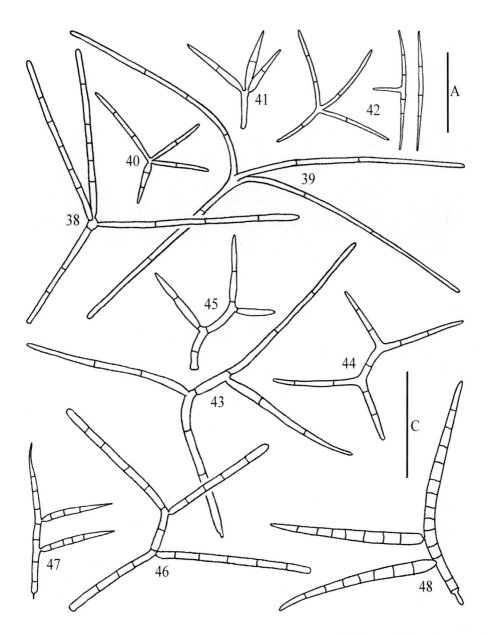

Figures 21.38—48. Conidia of aquatic hyphomycetes. 38. Geniculospora inflata. *39.*
Tetrachaetum elegans. *40.* Stenocladiella neglecta. *41.* Alatospora pulchella. *42.* Alatospora
acuminata. *43.* Tricladium chaetocladium. *44.* Tricladium angulatum. *45.* Tricladium
curvisporum. *46.* Tricladium patulum. *47.* Tricladium attenuatum. *48.* Tricladium splendens.
Scale bar A (Figs. 40—42, 45) = 25 μm, C (Figs. 38, 39, 43, 44, 46—48) = 50 μm.

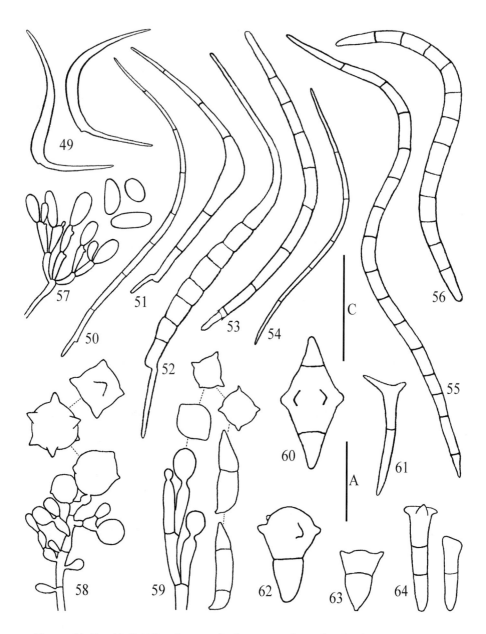

Figures 21.49—64. Conidia of aquatic hyphomycetes. Some details of conidiogenesis are shown in Figs. 57—59. 49. Lunulospora curvula. 50. Anguillospora filiformis. 51. Mycofalcella calcarata. 52. Mycocentrospora acerina. 53. Anguillospora longissima. 54. Flagellospora curvula. 55. Anguillospora furtiva. 56. Anguillospora crassa. 57. Dimorphospora foliicola. 58. Goniopila monticola. 59. Margaritispora aquatica. 60. Tumularia aquatica. 61. Heliscus submersus. 62. Tumularia tuberculata. 63. Heliscina campanulata. 64. Heliscus lugdunensis. Scale bar A (Figs. 58—64) = 25 µm, C (Figs. 49—57) = 50 µm.

5. REFERENCES

Bärlocher, F. (1992). *The Ecology of Aquatic Hyphomycetes*. Springer-Verlag. Berlin.

Bärlocher, F. (2004). Freshwater fungal communities. In: J. Dighton, P. Oudemans & J. White (eds.), *The Fungal Community*. 3rd ed. Marcel Dekker. New York (in press).

Bärlocher, F. & Marvanová, L. (2005). Aquatic hyphomycetes. In: D. F. McAlpine & I. M. Smith (eds.) *Biodiversity in the Atlantic Maritime Ecozone*. NRC Research Press. Ottawa (in press).

Descals, E., Marvanová, L. & Webster, J. (1998). New taxa and combinations of aquatic hyphomycetes. *Canadian Journal of Botany, 76*, 1647–1659.

Descals, E. & Webster, J. (1980). Taxonomic studies on aquatic hyphomycetes II. The *Dendrospora* aggregate. *Transactions of the British Mycological Society*, 74, 135–158.

Dudka, I.O. (1974). *Aquatic hyphomycetes of the Ukraine*. Naukova Dumka. Kiev. [In Ukrainian].

Gessner, M.O., Bärlocher, F. & Chauvet, E. (2003). Qualitative and quantitative analyses of aquatic hyphomycetes in streams. In: C.K.M. Tsui & K.D. Hyde (eds.), *Freshwater Mycology* (pp. 127–157). Fungal Diversity Press. Hong Kong.

Hieber, M. & Gessner, M.O. (2002). Contribution of stream detritivores, fungi, and bacteria to leaf breakdown based on biomass estimates. *Ecology, 83,* 1026–1038.

Ingold, C. T. (1975). *An illustrated Guide to Aquatic and Water-borne Hyphomycetes (Fungi Imperfecti) With Notes on their Biology*. Freshwater Biological Association Scientific Publication No. 30. Ambleside.

Marvanová, L. (1975). Concerning *Gyoerffyella* Kol. *Transactions of the British Mycological Society*, 65, 555–565.

Marvanová, L. (1997). Freshwater hyphomycetes: A survey with remarks on tropical taxa. In: K.K. Janardhanan, C. Rajendran, K. Natarajan & D.L. Hawksworth (eds.), *Tropical Mycology* (pp. 169–226). Science Publishers. Enfield.

Nilsson, S. (1964). Freshwater hyphomycetes. Taxonomy, morphology and ecology. *Symbolae Botanicae Upsaliensis, 18,* 1–130.

Petersen, R.H. (1962). Aquatic hyphomycetes from North America. I. Aleuriosporae (part 1) and key to the genera. *Mycologia, 54*, 117–151.

Petersen, R.H. (1963a). Aquatic hyphomycetes from North America. II. Aleuriosporae (part 2), and Blastosporae. *Mycologia*, 55, 18–29.

Petersen, R.H. (1963b). Aquatic hyphomycetes from North America III. Phialosporae and miscellaneous species. *Mycologia*, 55, 570–581.

Roldán, A., Descals, E. & Honrubia, M. (1989). Pure culture studies of *Tetracladium*. *Mycological Research*, 93, 452–465.

Suberkropp, K. (1991). Relationships between growth and sporulation of aquatic hyphomycetes on decomposing leaf litter. *Mycological Research*, 95, 843–850.

Suberkropp, K. & Klug, M. J. (1976). Fungi and bacteria associated with leaves during processing in a woodland stream. *Ecology*, 57, 707–719.

Webster, J. & Descals, E. (1981). Morphology, distribution, and ecology of conidial fungi in freshwater habitats. In G.T. Cole & B. Kendrick (eds.), *Biology of Conidial Fungi*, vol. 1 (pp. 295–355). Academic Press. New York.

CHAPTER 22

MOLECULAR APPROACHES TO ESTIMATE FUNGAL DIVERSITY. I. TERMINAL RESTRICTION FRAGMENT LENGTH POLYMORPHISM (T-RFLP)

LILIYA G. NIKOLCHEVA & FELIX BÄRLOCHER

63B York Street, Dept. of Biology, Mt. Allison University, Sackville, NB, Canada E4L 1G7.

1. INTRODUCTION

The major ecological function of aquatic hyphomycetes centres around the breakdown of leaves and other plant detritus in streams and rivers (Bärlocher 1992, Gessner et al. 2003). To fully characterize the fungal contribution to leaf decay and invertebrate nutrition, it is essential to subdivide the community into its constituent species and to determine the contribution of each species to the total fungal biomass production and leaf breakdown.

The most commonly used approach to describe community structure of aquatic hyphomycetes is based on inducing sporulation in mycelia present in the substrate (Gessner et al. 2003, Chapter 24). Released conidia are captured on a membrane filter, stained, counted and identified. This method will miss mycelia that do not sporulate, either because they are too small (recently established), too old (e.g. because they are dormant), or too slow to form conidia during standard incubation periods (commonly 48 h). Among aquatic hyphomycetes, the number of released spores is not necessarily correlated with mycelial biomass on the substrate (Bermingham et al. 1997), and in some cases reliable identification may be hampered by broadly overlapping spore morphologies (Chapter 21). The conventional approach will also miss the presence of fungal taxa other than aquatic hyphomycetes (e.g., some Ascomycota and Basidiomycota. Chytridiomycota, Zygomycota, and Oomycota).

Molecular approaches, based on characterizing the sequence and diversity of nucleic acids (most notably DNA), potentially circumvent these problems. For example, a recently developed system for field use can detect and identify a bacterial species in less than 10 min (Belgrader et al. 1999). The advantage of molecular

M.A.S. Graça, F. Bärlocher & M.O. Gessner (eds.), Methods to Study Litter Decomposition:
A Practical Guide, 169 – 176.

techniques in general is their extreme sensitivity (very low microbial biomass can be detected), and their applicability to all stages of the microbial life cycle.

When applied to microbial community diversity, molecular techniques most often rely on the amplification of DNA with taxon-specific primers (Borneman & Hatrin 2000) and subsequent characterization of the diversity of amplified DNA (Head et al. 1998). A high-throughput technique developed in bacterial ecology is terminal restriction fragment length polymorphism (T-RFLP). In T-RFLP, the extracted DNA is amplified with one or both of two primers fluorescently labelled at the 5' end. The products of the polymerase chain reaction (PCR) are then digested with a restriction enzyme, and the labelled terminal fragments are separated and detected on a DNA sequencer. The number of DNA fragments of different sizes gives an estimate of the minimum number of strains present in the analyzed community (Liu et al. 1997). Thus, the number of fragments detected is indicative of the number of phylogenetically different strains (phylotypes) present in a sample. The pattern of fragments obtained from T-RFLP is compared to the fragment lengths generated from digestion of DNA from pure cultures of aquatic hyphomycetes (Nikolcheva et al. 2003). Although this technique cannot be used to identify fungal species or strains, it provides a good estimate of the number of fungal phylotypes which is related to the fungal species richness on each substrate. This fast, high-throughput technique has been used to characterize fungi associated with *Spartina alterniflora* decaying in salt marshes (Buchan et al. 2002), and with soil fungi (Klamer et al. 2002).

The method for T-RFLP presented here is modified from Nikolcheva et al. (2003) and describes DNA extraction, amplification and purification, followed by endonuclease digestion and separation of the fragments. The technique has been adapted to the specific materials and equipment available at our laboratory but can easily be modified for other laboratory set-ups.

2. EQUIPMENT, CHEMICALS AND SOLUTIONS

2.1. Equipment and Material

- Pure fungal cultures
- Autumn-shed leaves, air dried
- Litter bags (10 x 10 cm, 10, 1 or 0.5 mm mesh size)
- Filtering apparatus and 8 μm membrane filters (e.g. Millipore Corporation, Bedford, MA)
- Freeze-dryer
- Mortar and pestle
- MoBio Ultra Clean Soil DNA extraction kit (MoBio, Solana Beach, CA, USA) or Nucleon PhytoPure Plant DNA Extraction Kit (Amersham Biosciences, Piscataway, NJ, USA)
- 3 pipettes: ranges of 0.5—10, 2—20 and 20—200 μl
- Pipette tips: 0.1—20 μl (white) and 2—200 μl (yellow)

- Ready-to-go PCR Beads (Amersham Biosciences, Piscataway, New Jersey, USA). Any other available PCR mix should work as well.
- Chambers for running horizontal agarose gels
- PCR thermocycler
- Parafilm
- DNA automated sequencer with fragment analysis software (e.g. Visible Genetics Long Read Tower by Visible Genetics, Suwanee, Georgia, USA)
- Water bath at 37 °C
- Ice bath

2.2. Chemicals

- Agarose (molecular biology grade)
- Malt broth
- Liquid nitrogen
- Tris(hydroxymethyl)aminomethane (TRIS, analytical grade)
- Ethylenediaminetetraaceticacid (EDTA, analytical grade)
- Sodium acetate
- Boric acid
- Acrylamide (6% for sequencing gels, Visible Genetics). The acrylamide must be compatible with the type of automated sequencer used.
- Bromophenol blue
- Glycerol
- Autoclaved deionized water (e.g. Milli-Q quality)
- Primers for the PCR reaction, specifically the fungal-specific primer F1300 (5' GATAACGAACGAGACCTTAAC 3'; Nikolcheva et al. 2003) labelled at the 5' end with Cy5.5 and the primer D (5' CYGCAGGTTCACCTAC 3'; Elwood et al. 1985) labelled at the 5' end with Cy5. Primer NS8 (5' TCCGCAGGTTCACCTACGGA 3'; White et al. 1990) can be used instead of primer D.
- Fluorescent DNA stain, preferably GelStar (BioWhittacker Molecular Applications, Rockland, Maine, USA), but ethidium bromide or other stains can also be used.
- DNA ladder with 100 or 250 base pairs (bp; Amersham Biosciences)
- Dimethyl sulfoxide (DMSO, analytical grade)
- Formamide loading dye, which is included with the sequencing kit; it keeps the DNA single-stranded before separation of the fragments by gel electrophoresis.
- GFX PCR and gel band purification kit (Amersham Biosciences)
- Restriction enzymes CfoI, RsaI, HinfI and DdeI (Roche Diagnostics, Indianapolis, IN, USA)
- Buffers for restriction enzymes: SuRE cut buffer L and buffer H (Roche Diagnostics); these buffers are supplied with the restriction enzymes.

- ALF Express Sizer 50—500 bases (Amersham Biosciences). This DNA standard works only if the automated sequencer can detect Cy5 and Cy5.5 fluorescent dyes. If the sequencer has a different detection system, DNA fragments of standard sizes labelled with appropriate dyes must be used.

2.3. Solutions

- 5× TBE (450 mM Tris, 450 mM boric acid, 13 mM EDTA)
- 1× TBE (100 ml 5× TBE and 400 ml deionized water)
- 0.5× TBE (100 ml 5× TBE and 900 ml deionized water)
- Agarose gel loading buffer (5 ml glycerol, 5 ml water, 5 mg bromophenol blue)
- 100× GelStar DNA stain (2 µl concentrated GelStar, 198 µl DMSO)

3. EXPERIMENTAL PROCEDURES

3.1. Sample Preparation

1. Grow pure fungal cultures in malt broth for 3 weeks. Filter through an 8-µm filter and freeze-dry overnight. The collected mycelia can be stored indefinitely at –80 °C.
2. Expose leaves in litter bags in the field and collected as described in Chapter 6.
3. Rinse leaves to remove any adhering particles, freeze-dry and store at –80 °C.

3.2. DNA Extraction

1. Use 30 mg of the dry pure fungal culture or up to 50 mg of dry plant tissue. Gently grind the samples with a mortar and pestle in liquid nitrogen for no more than 1 min.
2. Use the MoBio UltraClean Soil DNA extraction kit and follow the manufacturer's instruction for DNA isolation. The Nucleon PhytoPure Kit may also be used. Elute or resuspend the extracted DNA in 50 µl autoclaved deionized water.
3. The extracted DNA can be stored for up to 2 months at 4 °C or indefinitely at –20 °C.

3.3. DNA Amplification

1. To a Ready-to-go PCR bead add 19 µl of deionized autoclaved water, 2 µl of 10 µM forward primer (F1300; Nikolcheva et al. 2003), 2 µl of 10 µM reverse primer (D; Elwood et al. 1985) and 2 µl of extracted DNA.
2. Place all amplification reactions in a thermocycler programmed for 2 min of 95 °C initial denaturation followed by 35 cycles of denaturation at 95 °C for 30 s, primer annealing at 55 °C for 30 s and primer extension at 72 °C for 1 min.

End the program with 72 °C for 5 min of final extension. Store the PCR products at 4 °C.

3.4. Agarose Gel Electrophoresis

1. To check the concentration and quality of the PCR products, run an agarose gel.
2. Mix 30 ml 0.5× TBE buffer with 300 mg agarose. Microwave until the agarose has melted, then cast the gel. Let it solidify for 20 min.
3. Mix 1 μl loading dye, 0.7 μl 100× GelStar and 2 μl of PCR product (or 0.5 μg DNA ladder) on a piece of Parafilm.
4. Load the entire reaction in a well of the gel. Run the gel in 0.5× TBE buffer at 160 V for 15 min.
5. View on a UV transilluminator (wear protective glasses!).

3.5. DNA Purification

1. Purify the PCR products from solution (there should be 23 μl left) with GFX DNA and the gel band purification kit. Use the manufacturer's protocol for purification of PCR products from solution.
2. If the calculated DNA concentration from the agarose gel is between 0.1 and 0.3 μg ml^{-1}, elute the DNA with 50 μl deionized water. If the concentration is lower, elute in 25 μl.

3.6. Restriction Digest

1. Each digestion reaction contains 4.5 μl purified PCR product, 0.5 μl 10× enzyme buffer (buffer L is suitable for CfoI and RsaI, buffer H is suitable for DdeI and HinfI) and 5 U restriction enzyme.
2. If the restriction enzyme cleavage of DNA fragments is performed with a single enzyme, use 5 U of that enzyme. If the restriction is performed with two enzymes acting simultaneously, use 2.5 U of each. The enzymes CfoI and RsaI are compatible with each other, because both require SuRe cut buffer L; DdeI and HinfI are also compatible with one another, as they require SuRe cut buffer H.
3. Incubate the digestion reactions in a water bath or in the thermocycler at 37 °C for 2 h. The digestion products should be used immediately or stored at –20 °C.

3.7. Separation of Terminal Rrestriction Fragments

1. Prepare digestion products for loading on sequencing gel or capillary.
2. Combine 2 μl digestion product, 3 μl deionized water and 3 μl formamide loading dye.
3. For each set of 8 reaction mixtures, use 2 μl of the ALF express 50—500 bp sizer and 3 μl loading dye to run an external control.

4. Incubate all mixtures in a preheated thermocycler at 90 °C for 3 min to denature the DNA and then place them immediately on ice to keep the DNA single-stranded.

5. Load 2 μl of each sample on a sequencing gel within 1 h after denaturation.

6. Run the gel in 1× TBE buffer at 1500 V and 52 °C for 45 min. The conditions for electrophoresis are specific to the automated sequencer used.

4. FINAL REMARKS

For calibration, the DNA from pure fungal cultures can be analyzed before the environmental samples. Each fungal species should theoretically yield one fragment, and ideally, the fragment length is species-specific. However, some primers will amplify conserved genes and some enzymes will cut at a conserved region in DNA; when this happens, fragments of different species or genera may be identical. The results of a T-RFLP analysis (i.e., number of fragments differing in length) therefore provide an estimate of how much variability can be detected with a given primer/enzyme combination, and true diversity will generally be underestimated. In our experience, the enzyme DdeI gives the highest variability among pure fungal cultures. This restriction enzyme was therefore used to analyze environmental samples (Fig. 22.1). The fragment lengths obtained from pure cultures can be compared to observed lengths in environmental samples; if the peaks on the chromatograms coincide, the environmental samples may contain the particular fungal species. However, the fragments in the environmental samples cannot be identified with certainty.

Rather than using extraction kits, DNA can be extracted from samples with a standard SDS or CTAB-based phenol-chloroform procedure. The DNA yield from these types of extractions can be very high, but in our experience the DNA quality is often poor, making amplification by PCR impossible.

In our experience, the 18S rRNA restrictions enzymes CfoI and DdeI detect the highest community variability (Nikolcheva et al. 2003), and a combination of restriction enzymes may result in an even higher interspecific variability (Nikolcheva & Bärlocher 2004).

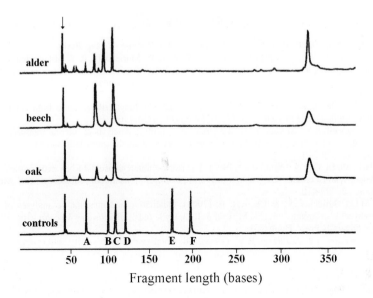

Fragment length (bases)

Fig. 22.1. T-RFLP chromatograms from fungal communities associated with alder, beech, and oak leaves after 7 days of exposure in a stream. The 3' end of the 18S rRNA gene was amplified with a fungal-specific primer pair and digested with DdeI. The peak designated by the arrow is unextended primer. The lane labeled "controls" contains fragments from the digestion of A: Anguillospora furtiva, *B:* Colispora elongata, *C:* Articulospora tetracladia, *D:* Anguillospora rubescens, *E:* Tetracladium marchalianum, *and F:* Heliscus lugdunensis.

5. REFERENCES

Bärlocher, F. (1992). *The Ecology of Aquatic Hyphomycetes.* Springer-Verlag. Berlin.

Belgrader, P., Benett, W., Hadley, D., Richards, J., Stratton, P., Mariella Jr., R. & Milanovich, F. (1999). PCR detection of bacteria in seven minutes. *Science*, 284, 449-450.

Bermingham, S., Maltby, L. & Dewey, F.M. (1997). Use of immunoassays for the study of natural assemblages of aquatic hyphomycetes. *Microbial Ecology*, 33, 223-229.

Borneman, J. & Hatrin, R.J. (2000). PCR primers that amplify fungal rRNA genes from environmental samples. *Applied and Environmental Microbiology*, 66, 4356-4360.

Buchan, A., Newell, S.Y., Moreta, J.I.L. & Moran, M.A. (2002). Analysis of internal transcribed spacer (ITS) regions of rRNA genes in fungal communities in a southeastern U.S. salt marsh. *Microbial Ecology*, 43, 329-340.

Elwood, H.J., Olsen, G.J. & Sogin, M.L. (1985). The small-subunit ribosomal RNA gene sequences from the hypotrichous ciliates *Oxytricha nova* and *Stylonychia pustulata*. *Molecular and Biological Evolution*, 2, 399-410.

Gessner, M.O., Bärlocher, F. & Chauvet, E. (2003). Qualitative and quantitative analyses of aquatic hyphomycetes in streams. In: C.K.M. Tsui & K.D. Hyde (eds.). *Freshwater Mycology* (pp. 127-157). Fungal Diversity Press. Hong Kong.

Head, I.M., Sanders, J.R. & Pickup, R.W. (1998). Microbial evolution, diversity, and ecology: a decade of ribosomal RNA analysis of uncultivated organisms. *Microbial Ecology*, 35, 1-21.

Klamer, M., Roberts, M.S., Levine, L.H., Drake, B.G. & Graland, J.L. (2002). Influence of elevated CO_2 on the fungal community in a coastal scrub oak forest soil investigated with Terminal-Restriction Fragment Length Polymorphism analysis. *Applied and Environmental Microbiology*, 68, 4370-4376.

Liu, W.-T., Marsh, T.L., Cheng, H. & Forney, L.J. (1997). Characterization of microbial diversity by determining terminal restriction fragment length polymorphisms of genes encoding 16S rRNA. *Applied and Environmental Microbiology*, 63, 4516-4522.

Nikolcheva, L.G. & Bärlocher,F. (2004). Seasonal and substrate preferences of fungi colonizing leaves in streams: traditional vs. molecular evidence. *Environmental Microbiology*, in press.

Nikolcheva, L.G., Cockshutt, A.M. & Bärlocher, F. (2003). Diversity of freshwater fungi on decaying leaves: comparing traditional and molecular approaches. *Applied and Environmental Microbiology*, 69, 2548-2554.

White, T.J., Bruns, T., Lee, S. & Taylor, J.W. (1990). Amplification and direct sequencing of fungal ribosomal RNA genes for phylogenetics. In: M. A. Innis, D. H. Gelfand, J. J. Sninsky & T.J. White (eds.). *PCR protocols: A guide to Methods and Applications* (pp. 315-322). Academic Press, Inc. New York.

CHAPTER 23

MOLECULAR APPROACHES TO ESTIMATE FUNGAL DIVERSITY. II. DENATURING GRADIENT GEL ELECTROPHORESIS (DGGE)

LILIYA G. NIKOLCHEVA & FELIX BÄRLOCHER

63B York Street, Dept. of Biology, Mt. Allison University, Sackville, NB, Canada E4L 1G7.

1. INTRODUCTION

In addition to T-RFLP (Chapter 22), a range of other molecular methods have been developed to determine the species composition of microbial communities. For example, taxon-specific primers are used to amplify a gene of interest (Borneman & Hatrin 2000), which is then cloned into a bacterial vector. Each bacterial clone is grown on a plate to produce an individual colony, and the cloned gene from each colony is sequenced (Head et al. 1998). The sequence is then compared to sequences published in a genomic database such as GenBank (http://www.ncbi.nlm.nih.gov/). This technique accurately determines the identity and phylogeny of microorganisms in diverse communities. However, it is both time-consuming and expensive. Moreover, since clones are randomly sampled before sequencing, it is likely that some clones are never sampled, and the diversity of the community is thus underestimated. The cloning and sequencing approach has been used for analyzing natural fungal communities associated with plant roots in soil (Vandenkoornhuyse et al. 2002) and with decomposing leaves of the salt-marsh grass, *Spartina alterniflora* (Buchan et al. 2002).

Denaturing gradient gel electrophoresis (DGGE) combines the advantages of cloning and sequencing and T-RFLP. A gene of interest present in all members of a community, such as the aquatic hyphomycete community on a leaf, is targeted with appropriate primers and amplified through PCR. If the genes of the various species differ significantly in length, they can be separated by conventional electrophoresis. However, species- or strain-specific differences may be more subtle: they may involve a slight change in base composition (AT versus GC), while maintaining the same overall length. Double-stranded DNA sequences of identical lengths but of

M.A.S. Graça, F. Bärlocher & M.O. Gessner (eds.), Methods to Study Litter Decomposition:
A Practical Guide, 177 – 184.

different base composition cannot normally be separated by electrophoresis; however, they differ in their ease of denaturation (i.e. separation of doubled-stranded DNA into single strands). DGGE exploits this difference to separate amplified DNA on a high-resolution polyacrylamide gel along a denaturing gradient (Fisher & Lerman 1983, Muyzer et al. 1993). Sequences differing in base composition denature at different locations on this gel. When denaturation is initiated, a 'bubble' forms at the site where the two strands separate. This dramatically slows down migration of the DNA through the gel. The number of bands on the gel is therefore indicative of the gene diversity in the original sample, such as a DNA extract from a decomposing leaf colonized by microorganisms.

The potential advantage of the DGGE technique over T-RFLP is that DNA from each separate band on the gel can be isolated and sequenced to identify all genes in the sample. The sequences can be compared to those published in GenBank, which, depending on the gene, allows identification at the level of phylum, genus or even species.

The method presented here is modified from Nikolcheva et al. (2003). The procedures have been adapted to the specific materials and equipment available at our laboratory, but can easily be modified for other laboratory set-ups. With the primers described here, the method does not discriminate among members of different fungal phyla (e.g., Ascomycota, Basidiomycota). To achieve this, taxon-specific primers have to be used (Nikolcheva & Bärlocher 2004).

2. EQUIPMENT, CHEMICALS AND SOLUTIONS

2.1. Equipment and Materials

- Pure fungal cultures
- Autumn-shed leaves, air dried
- Litter bags (10 x 10 cm, 10, 1 or 0.5 mm mesh size)
- Filtering apparatus and 8 μm membrane filters (e.g. Millipore Corporation, Bedford, MA)
- Freeze-dryer
- Mortar and pestle
- MoBio Ultra Clean Soil DNA extraction kit (MoBio, Solana Beach, California) or Nucleon PhytoPure Plant DNA Extraction Kit (Amersham Biosciences, Piscataway, New Jersey, USA)
- Automatic pipettes: ranges of 0.5—10, 2—20 and 20—200 μl
- Pipette tips: 0.1—20 μl (white) and 2 – 200 μl (yellow)
- DCode Mutation Detection System (BioRad, Hercules, California) with Model 475 Gradient Former (BioRad) or any other DGGE system
- Ready-to-go PCR Beads (Amersham Biosciences); any other available PCR mix should work as well.
- Chambers for running horizontal agarose gels
- Direct current (DC) power supply

- PCR thermocycler
- Parafilm
- Ice bath

2.2. Chemicals

- Agarose (molecular biology grade)
- Malt broth
- Liquid nitrogen
- Tris(hydroxymethyl)aminomethane (TRIS, analytical grade)
- Ethylenediaminetetraacetic acid (EDTA, analytical grade)
- Sodium acetate
- Boric acid
- Acrylamide (40%), 37.5:1 acrylamide:bis-acrylamide
- Formamide (100%, analytical grade)
- Urea (analytical grade)
- N, N, N, N'-Tetramethylethylene diamine (TEMED, analytical grade)
- Ammonium persulfate (analytical grade)
- Bromophenol blue
- Glycerol
- Autoclaved deionized water (e.g., Milli-Q quality)
- Primers for the PCR reaction: NS1 (5' GTAGTCATATGCTTGTCTC 3', White et al. 1990) and GCfung sequence (May et al. 2001)
- Fluorescent DNA stain, preferably GelStar (BioWhittacker Molecular Applications, Rockland, Maine, USA) but ethidium bromide or other DNA stains can also be used
- DNA ladder with 100 or 250 base pairs (bp; Amersham Biosciences)
- Dimethyl sulfoxide (DMSO)

2.3. Solutions

- 5× TBE (450 mM Tris, 450 mM boric acid, 13 mM EDTA)
- 0.5× TBE (100 ml 5× TBE and 900 ml water)
- 50× TAE (2.0 M Tris, 1.0 M acetic acid, 50 mM EDTA, pH 8.3)
- 1× TAE (20 ml 50× TAE and 980 ml water)
- 8% acrylamide, 0% denaturant (20 ml 40% acrylamide, 2 ml 50× TAE, 78 ml water)
- 8% acrylamide, 100% denaturant (20 ml 40% acryl amide, 2 ml 50× TAE, 40 ml formamide, 42 g urea, 8 ml water)
- Agarose gel loading buffer (5 ml glycerol, 5 ml water, 5 mg bromophenol blue)
- 100× GelStar DNA stain (2 μl concentrated GelStar, 198 μl DMSO)

3. EXPERIMENTAL PROCEDURES

3.1. Sample Preparation

1. Grow pure fungal cultures in malt broth for 3 weeks. Filter through an 8 μm filter and freeze dry overnight. These mycelia can be stored indefinitely in a –80 °C freezer.
2. Expose leaves in litter bags in the field and collected as in Chapters 5 and 6.
3. Rinse leaves to remove any adhering particles, freeze-dry and store at –80 °C.

3.2. DNA Extraction

1. Use 30 mg of the dry pure fungal culture or up to 50 mg of dry plant tissue. Gently grind the samples with a mortar and pestle in liquid nitrogen for up to 1 min.
2. Use the MoBio UltraClean Soil DNA extraction kit and follow the manufacturer's instruction for DNA isolation. The Nucleon PhytoPure DNA Extraction kit can also be used. Elute or resuspend the DNA in 50 μl autoclaved deionized water. In our experience, SDS or CTAB-based phenol-chloroform DNA extraction can give very high yields, but the quality of DNA is poor, making amplification by PCR impossible.
3. The extracted DNA can be stored for up to 2 months at 4 °C or indefinitely at –20 °C.

3.3. DNA Amplification

1. To a Ready-to-go PCR bead add 19 μl of autoclaved deionized water, 2 μl of 10 μM forward primer (NS1; White et al. 1990), 2 μl of 10 μM reverse primer (GCfung; May et al. 2001) and 2 μl extracted DNA.
2. Place all amplification reactions in a thermocycler programmed for 2 min of 95 °C initial denaturation followed by 35 cycles of denaturation at 95 °C for 30 s, primer annealing at 55 °C for 30 s and primer extension at 72 °C for 1 min. End the program with 72 °C for 5 min of final extension. Store the PCR products at 4 °C.

3.4. Agarose Gel Electrophoresis

1. To check the concentration and quality of the PCR products, run an agarose gel.
2. Mix 30 ml 0.5× TBE buffer with 300 mg agarose. Microwave until the agarose has melted, then cast the gel. Let it solidify for 20 min.
3. Mix 1 μl loading dye, 0.7 μl 100× GelStar and 2 μl of PCR product (or 0.5 μg DNA ladder) on a piece of Parafilm.
4. Load the entire reaction in a well of the gel. Run the gel in 0.5× TBE buffer at 160 V for 15 min.

5. View on a UV transilluminator (wear protective glasses!).

3.5. Denaturing Gradient Gel Electrophoresis (DGGE)

1. Pour 7 l of 1× TAE buffer in the gel running chamber and start pre-warming to 56 °C. Depending on the room temperature, this will take 45—60 min.
2. Clean all glass plates with ethanol and assemble them with 1-mm spacers according to the manufacturer's instructions for parallel DGGE.
3. In a 50-ml plastic tube prepare 16 ml of a low-denaturant solution containing 20% denaturing gradient (100% denaturing gradient corresponds to 7 M urea and 40% formamide). To make the low-denaturant solution, mix 12.8 ml of 0% denaturant solution with 3.2 ml of 100% denaturant solution. In a different tube mix 7.2 ml of 0% denaturant solution and 8.8 ml of 100% denaturant solution (this makes 55% denaturant). Keep both solutions on ice.
4. Optionally add 60 μl of gel dye to the high-denaturant solution and mix. To both high- and low-denaturant solutions add 14.4 μl TEMED and 144 μl 10% APS and mix. The solutions will take about 10 min to polymerize if kept on ice; once they are at room temperature, they polymerize more quickly. All steps involving the casting of the gel should be performed as quickly as possible.
5. Fill two 30-ml syringes with tubing with the solutions, place on the Gradient Former and assemble the Tygon tubing as described by the manufacturer. At the end of the tubing add an 18-gauge needle.
6. Cast the gel. The blue high-denaturant solution should be on the bottom and should form a uniform transition into the transparent low-denaturant solution on the top. Add a 16-well comb and let the gel polymerize for 45 min at room temperature.
7. If a second gel is made, wash the plastic tubes, syringes and tubing before repeating all steps.
8. Once the gels have polymerized, assemble the gel running gasket according to the manufacturer's instructions. Place the gasket in the gel running chamber and pre-warm the gel for 10 min.
9. Prepare the samples for loading. Mix 12 μl PCR product with 12 μl gel loading dye. Load lanes 2—15 with a yellow 2—200 μl pipette tip. The outer lanes on the gel do not run at a consistent rate, thus forming crescent-shaped rather than straight DNA bands.
10. Turn on the pump and the heater; attach the gel running apparatus to a DC power supply. There can be slight fluctuations in the current during the run, probably due to temperature fluctuations. In our experience, an older, less sensitive power supply works more consistently than an electronic power supply.
11. Run the gel at 50 V and 56 °C for 16 h. The voltage can be increased up to 80 V, and the running time should then be decreased accordingly to 9 h.
12. The gel dye and the loading dye migrate out of the gel and into the buffer during the run. The gel should be transparent. Switch off the power supply, the

pump and the heater. Take the gel gasket out of the chamber and disassemble the gels. Prepare staining solution for each gel: 30 ml of 1× TAE with 3 µl 10,000× GelStar. Stain each gel in the staining solution for 10 min.

13. View the gels on a UV transilluminator (wear protective glasses!).

14. Analyze the band intensity in each lane using image analysis software (e.g. National Institutes of Health, Bethesda, Maryland, USA).

15. Initially, analyze DNA from pure fungal cultures to determine the variability in the sequences and to generate a set of standards that can be used to calibrate environmental samples (Fig. 23.1).

16. Use a mixture of PCR products from pure fungal cultures whose separation is optimized as a standard on a gel with environmental samples (Fig. 23.2). Some bands from an environmental sample may migrate at the same rate as bands from pure fungal cultures, which suggests that a particular fungal species may be present in the environmental sample. The DNA from the bands should then be extracted and sequenced to confirm the identity of the species.

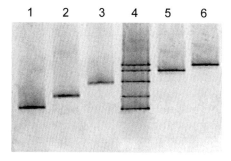

Fig. 23.1. Amplified DNA from pure cultures of aquatic hyphomycetes separated by DGGE. Lane 1: Tumularia aquatica; *2:* Articulospora tetracladia; *3:* Tetracladium marchalianum; *4: Mix of all five cultures; 5:* Heliscus lugdunensis; *6:* Anguillospora furtiva.

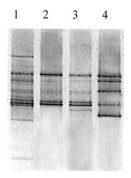

Fig. 23. 2. Amplified fungal DNA from environmental samples separated by DGGE. Leaves of alder, beech and oak were submerged in a stream for 7 days, the DNA was extracted, amplified and separated. Lane 1: fungal DNA from alder leaves, 2: beech leaves, 3: oak

leaves, 4: amplification products from pure cultures of aquatic hyphomycetes used as standards (same species as in Fig. 23.1).

4. REFERENCES

Borneman, J, & Hatrin, R.J. (2000). PCR primers that amplify fungal rRNA genes from environmental samples. *Applied and Environmental Microbiology*, 66, 4356-4360.

Buchan, A., Newell, S.Y., Moreta, J.I.L. & Moran, M.A. (2002). Analysis of internal transcribed spacer (ITS) regions of rRNA genes in fungal communities in a southeastern U.S. salt marsh. *Microbial Ecology*, 43, 329-340.

Fischer, S.G. & Lerman, L.S. (1983). DNA fragments differing by single base-pair substitutions are separated in denaturing gradient gels: correspondence with melting theory. *Proceedings of the National Academy of Sciences, USA*, 80, 1579-1583.

Head, I.M., Sanders, J.R. & Pickup, R.W. (1998). Microbial evolution, diversity, and ecology: a decade of ribosomal RNA analysis of uncultivated organisms. *Microbial Ecology*, 35, 1-21.

May, L.A., Smiley, B. & Schmidt, M.G. (2001). Comparative denaturing gradient gel electrophoresis of fungal communities associated with whole plant corn silage. *Canadian Journal of Microbiology*, 47, 829-841.

Muyzer, G., de Waal, E.C. & Uitterlinden, A.G. (1993). Profiling of complex microbial populations by denaturing gradient gel electrophoresis analysis of polymerase chain reaction – amplified genes coding for 16S rRNA. *Applied and Environmental Microbiology*, 59, 695-700.

Nikolcheva, L.G. & Bärlocher, F. (2004) Taxon-specific fungal primers reveal unexpectedly high diversity during leaf decomposition in a stream. *Mycological Progress*, 3, 41-50.

Nikolcheva, L.G., Cockshutt, A. M. & Bärlocher, F. (2003). Diversity of freshwater fungi on decaying leaves: comparing traditional and molecular approaches. *Applied and Environmental Microbiology*, 69, 2548-2554.

Vandenkoornhuyse, P., Baldauf, S.L., Leyval, C., Straczek, J. & Young, J.P.W. (2002). Extensive fungal diversity in plant roots. *Science*, 295, 2051.

White, T.J., Bruns, T., Lee, S. & Taylor, J.W. (1990). Amplification and direct sequencing of fungal ribosomal RNA genes for phylogenetics. In M. A. Innis, D. H. Gelfand, J. J. Sninsky & T.J. White (eds.), *PCR Protocols: A Guide to Methods and Applications* (pp. 315-322). Academic Press. New York.

CHAPTER 24

SPORULATION BY AQUATIC HYPHOMYCETES

FELIX BÄRLOCHER

63B York Street, Dept. of Biology, Mt. Allison University, Sackville, NB, Canada E4L 1G7.

1. INTRODUCTION

Fungi are instrumental in leaf decomposition in streams (Gessner & Chauvet 1994), and their biomass accumulating on leaves improves substrate palatability and nutritional value to shredders (Bärlocher 1985, Suberkropp 1992, Graça 1993, 2001). The preferred method to measure fungal biomass is based on the indicator molecule ergosterol, which occurs at a relatively constant concentration in living mycelia (Chapter 25). A very substantial proportion of fungal production, often in excess of 50% (Findlay & Arsuffi 1989, Chauvet & Suberkropp 1998, Sridhar & Bärlocher 2000), is invested in propagules that are released from leaves. Asexually produced spores (mitospores, conidia) dominate. On leaves freshly recovered from a stream, only a few conidia can be observed. However, if such leaves are incubated for 1—2 days under conditions that stimulate sporulation (low to intermediate nutrient levels, high turbulence), newly formed conidia will be released. They can be trapped on a membrane filter, stained, and counted and identified under a microscope. There is a significant correlation between maximum fungal biomass on the leaf and maximum spore production over the course of decomposition (Gessner & Chauvet 1994, Maharning & Bärlocher 1996). However, at any given point during decomposition, high sporulation rate by a species does not imply the presence of high mycelial biomass belonging to the same species on the leaf (Bermingham et al. 1997). Sporulation under laboratory conditions can be as high as 4000 spores produced day^{-1} mg^{-1} of leaf dry mass (for review, see Gessner 1997; selected values are shown in Table 24.1).

The aim of this chapter is an estimate of the reproductive potential of the mycelia present in leaves recovered from a stream, following procedures based on Bärlocher (1982). This and similar procedures are also described in Gessner et al. (2003). The data can be used to estimate the amount of conidial biomass released from leaves, or, to describe the diversity and composition of the fungal community.

M.A.S. Graça, F. Bärlocher & M.O. Gessner (eds.), Methods to Study Litter Decomposition:
A Practical Guide, 185 – 188.
© 2005 *Springer. Printed in The Netherlands.*

Table 24.1. Maximum spore production rates (no. day^{-1} mg^{-1} of leaf dry mass) reported from leaves decomposing in streams, from selected studies. Length = number of days of incubation in stream before maximum was reached.

Sporulation rate	Length	Leaf species	Condition	Reference
75	93	*Liriodendron tulipifera*	Softwater	1
425	21	*Liriodendron tulipifera*	Hardwater	1
1500	28	*Alnus glutinosa*	Softwater, 10 °C	2
4000	14	*Alnus glutinosa*	15 °C	3
7000	28	*Eucalyptus globulus*	15 °C	3

1 = Suberkropp (2001); 2 = Hieber & Gessner (2002); 3 = Bärlocher et al. (1995)

2. EQUIPMENT, CHEMICALS AND SOLUTIONS

2.1. Equipment and Material

- Autumn-shed leaves, air dried
- Litter bags (10 × 10 cm, 10, 1 or 0.5 mm mesh size).
- Erlenmeyer flasks, 250 ml, with 150 ml of deionized, sterile water
- Membrane filters, 5 μm pore size, and filtering apparatus
- Supply of pressurized air (e.g. aquarium pumps), tubing and Pasteur pipettes, or shaker
- Drying oven (40—50 °C)
- Balance (±1 mg precision)
- Microscope (with 16, 40 and 100× objectives)

2.2. Chemicals and Solutions

- Lactic acid
- Deionized water
- Phenol and glycerol for long-term storage
- 0.1% Trypan Blue or Cotton Blue in 60% lactic acid (Trypan Blue in lactophenol is preferable for long-term storage: 10 ml phenol, 10 ml lactic acid, 20 ml glycerol, 10 ml H$_2$O)

3. EXPERIMENTAL PROCEDURES

1. Prepare litter bags to be placed in a stream as indicated in Chapters 5 and 6.
2. Anchor leaf bags to stream bed by means of bricks, steel pegs or other devices. Be careful not to place too many bags close to each other, because this may greatly change flow patterns and thereby affect fungal colonization of leaves.
3. Recover bags at appropriate intervals (fungal colonization proceeds faster on leaf species that are more rapidly decomposed; see Chapter 6).
4. Rinse leaves to remove silt, sand and invertebrates.
5. Place some leaf material (ca. 9 cm^2) in an Erlenmeyer flask with sterile, deionized water or filtered stream water.

6. Induce turbulence by placing the flask on a shaker (100—150 rpm) or by aerating it (connect Pasteur pipettes with tubing to source of pressurized air and adjust air-flow to approx. 1 ml s^{-1}).
7. After 24—48 h, remove leaf material and determine its dry mass (40—50 °C, 2 days or until constant weight is reached).
8. Filter supernatant through membrane filter.
9. Add a few drops of Trypan Blue solution to filter; incubate for 30—60 min at 40—50 °C.
10. Scan the surface of the filter under the light microscope. Count and identify all conidia, or, if they are very numerous, all conidia in 20—30 randomly chosen microscope fields, or, on a defined fraction of the filter (cf. Gönczöl et al. 2001).
11. Express the number of spores produced during laboratory incubations per leaf dry mass or ash-free dry mass.
12. To estimate biomass of the spores, determine total volume of spores, and assume a density of 500 fg μm^{-3} (Findlay and Arsuffi 1989). Volumes of selected species are listed in Bärlocher & Schweizer (1983). Chauvet & Charcosset (2000) provide average spore masses of additional species. Or, assume an average conidial biomass of 200 pg (conservative estimate; Gessner 1997).
13. Analyze fungal community structure as described in Chapter 42 or by other means (e.g. multivariate analyses).

4. REFERENCES

Bärlocher, F. (1982). Conidium production from leaves and needles in four streams. *Canadian Journal of Botany*, 60, 1487-1494.

Bärlocher, F. (1985). The role of fungi in the nutrition of stream invertebrates. *Botanical Journal of the Linnean Society*, 91, 83-94.

Bärlocher, F., Canhoto, C. & Graça, M.A.S. (1995). Fungal colonization of alder and eucalypt leaves in two streams in Central Portugal. *Archiv für Hydrobiologie*, 133, 457-470

Bärlocher, F. & Schweizer, M. (1983). Effects of leaf size and decay rate on colonization by aquatic hyphomycetes. *Oikos*, 41, 205-210.

Bermingham, S., Maltby, L. & Dewey, F.M. (1997). Use of immunoassays for the study of natural assemblages of aquatic hyphomycetes. *Microbial Ecology*, 33, 223-229.

Chauvet, E. & Suberkropp, K. (1998). Temperature and sporulation of aquatic hyphomycetes. *Applied and Environmental Microbiology*, 64, 1522-1525.

Findlay, S.E.G. & Arsuffi, T.L. (1989). Microbial growth and detritus transformations during decomposition of leaf litter in a stream. *Freshwater Biology*, 21 261-269.

Gessner, M.O. (1997). Litter breakdown in rivers and streams. *Limnética*, 13, 33-44.

Gessner, M.O., Bärlocher, F. & Chauvet, E. (2003). Qualitative and quantitative analyses of aquatic hyphomycetes in streams. In: Tsui, C.K.M. & Hyde K.D. (eds.). *Freshwater Mycology* (pp. 120-157). Fungal Diversity Press. Hong Kong.

Gessner, M.O. & Chauvet, E. (1994). Importance of stream microfungi in controlling breakdown rates of leaf litter. *Ecology*, 75, 1807-1817.

Gönczöl, J., Révay, A. & Csontas, P. (2001). Effect of sample size on the detection of species and conidial numbers of aquatic hyphomycetes collected by membrane filtration. *Archiv für Hydrobiologie*, 150, 677-691.

Graça, M.A.S. (1993). Patterns and processes in detritus-based stream systems. *Limnologica*, 23, 107-114.

Graça, M.A.S. (2001). The role of invertebrates on leaf litter decomposition in streams - A review. *International Review of Hydrobiology*, 86, 383-393.

Hieber, M. & Gessner, M.O. (2002). Contribution of stream detritivores, fungi and bacteria to leaf breakdown based on biomass estimates. *Ecology*, 83, 1026-1038.

Maharning, A.R. & Bärlocher, F. (1996). Growth and reproduction in aquatic hyphomycetes. *Mycologia*, 88, 80-88.

Sridhar, K.R. & Bärlocher, F. (2000). Initial colonization, nutrient supply and fungal activity on leaves decaying in streams. *Applied and Environmental Microbiology*, 66, 1114-1119.

Suberkropp, K. (2001). Fungal growth, production, and sporulation during leaf decomposition in two streams. *Applied and Environmental Microbiology*, 67, 5063-5068

Suberkropp, K. (1992). Interactions with invertebrates. In: F. Bärlocher (Ed.), *The Ecology of Aquatic Hyphomycetes* (pp. 118-133). Ecological Studies, Vol. 94, Springer-Verlag, Berlin.

CHAPTER 25

ERGOSTEROL AS A MEASURE OF FUNGAL BIOMASS

MARK O. GESSNER

Department of Limnology, EAWAG, Limnological Research Center, 6047 Kastanienbaum, Switzerland.

1. INTRODUCTION

Fungi are an important component of decomposer assemblages associated with plant litter in streams and other environments. The biomass of fungi accumulating in decomposing litter can be substantial (Table 1), and this has been used as one line of evidence that fungi are instrumental in litter decomposition (Gessner & Chauvet 1994, Maharning & Bärlocher 1996). By colonizing and degrading litter, fungi also enhance its palatability to detritivorous invertebrates (shredders) and provide an attractive food source to these consumers (Suberkropp 1992, Graça 2001). Fungi thus play important roles in streams and other ecosystems relying on plant litter inputs. The method presented here is a means to assess their quantitative importance.

Determining fungal biomass in plant litter and other solid substrates has long proven difficult (Newell 1992, Gessner & Newell 2002), partly because fungal hyphae spread within their substrate rather than growing at surfaces. As a result, the fungal mycelium is not easily separated from the leaf tissue by either optical or mechanical methods. One way to circumvent this problem is to quantify a cell constituent that is specific to the target fungi and occurs in rather constant amounts in fungal mycelium. Chitin, a major cell wall component of eumycotic fungi, and ergosterol, a major membrane component, are the two main constituents that have been used to this end, but ergosterol appears to be superior if metabolically active biomass is to be determined (Newell 1992, Charcosset & Chauvet 2001, Gessner & Newell 2002).

The purpose of the method presented in this chapter is to determine the biomass of fungi in decomposing plant litter. However, the proposed procedure for extracting and quantifying ergosterol might also be used for gut contents and faeces of shredders that have been feeding on decomposing leaves or wood. Pure fungal mycelium may also be analyzed so as to establish conversion factors relating

M.A.S. Graça, F. Bärlocher & M.O. Gessner (eds.), Methods to Study Litter Decomposition: A Practical Guide, 189 – 196.

Table 25.1. Fungal biomass in decomposing leaves estimated by the ergosterol method. A conversion factor of 5.5 mg of ergosterol g^{-1} fungal biomass was used (Gessner & Chauvet 1993) except for the study by Gessner & Chauvet (1994), where species-specific conversion factors were applied.

Ergosterol ($\mu g\ g^{-1}$ AFDM)	Fungal biomass (% AFDM)	Peak time (day)	Leaf species	Stream type	Reference
437	6.6	56	*Fagus sylvatica*	Softwater, mountain	1*
631	9.8	28	*Alnus glutinosa*	Softwater, mountain	1*
477	8.7	55	*Alnus glutinosa*	Softwater, mountain	2*
316	5.8	45[†]	*Alnus viridis*	Softwater, glacial	3*
528	9.6	93	*Liriodendron tulipifera*	Softwater, lowland	4
704	12.8	57	*Liriodendron tulipifera*	Hardwater, lowland	4

* AFDM data not previously published. [†] Higher values noted in spring following snow melt.
1 = Gessner & Chauvet (1994); 2 = Hieber & Gessner (2002); 3 = Gessner et al. (1998); 4 = Suberkropp (2001).

ergosterol amounts to fungal mass. Determination of ergosterol is achieved by high-performance liquid chromatography (HPLC) after lipid extraction in alkaline methanol and purification of the extract by means of solid-phase extraction (SPE). The method is discussed in detail by Gessner & Newell (2002). The procedures adopted here have been slightly modified from Gessner & Schmitt (1996).

2. EQUIPMENT, CHEMICALS AND SOLUTIONS

2.1. Equipment and Material

- Screw-cap extraction tubes (approx. 40 ml, pressure-resistant)
- Water bath or dry bath
- Magnetic stirrer and stirring bars
- Vacuum manifold for solid phase extraction (SPE)
- Pump to create vacuum in manifold
- Solid-phase extraction cartridges (Waters Sep-Pak®, Vac RC, tC18, 500 mg sorbent)
- HPLC for isocratic operation (1 pump, injector, UV detector set to 282 nm, recording unit)
- HPLC column (e.g. LiChrospher RP18, 25 cm × 4.6 mm; Merck, Darmstadt; Germany)
- Gas-tight syringe (1 ml)
- Glassware, pipettes and plastic syringes

2.2. Chemicals

- Methanol (HPLC or analytical grade)
- Propanol-2 = isopropanol (HPLC or analytical grade)
- Ergosterol standard (purity > 98%; Fluka, Buchs, Switzerland)
- KOH (pellets, analytical grade)
- HCl (analytical grade)
- Boiling chips

2.3. Solutions

- Solution 1:   Methanol
- Solution 2:   Storage and extraction solvent: KOH in methanol: 8 g l^{-1} (e.g. 4 g in 0.5 l are sufficient for >30 samples)
- Solution 3:   0.65 M HCl (ca. 100 ml for 30 samples)
- Solution 4:   Conditioning solution: methanol (1 volumetric part) + KOH in methanol (5 parts) + 0.65 M HCl (1 part); e.g. 30 ml Solution 1 + 150 ml Solution 2 + 30 ml Solution 3; sufficient for >30 samples; check before use whether pH is < 3.
- Solution 5:   Washing solution: 0.4 M KOH in methanol:H$_2$O (6:4; vol:vol); e.g. 1.8 g KOH + 32 ml H$_2$O + 48 ml methanol; sufficient for > 30 samples
- Solution 6:   Isopropanol
- Solution 7:   Ergosterol standard in isopropanol: weigh ca. 10 mg ergosterol to nearest 0.1 mg in volumetric flask (50 ml), dissolve in isopropanol, adjust volume, and transfer to tightly closing 50-ml glass bottle; store in refrigerator (4 °C), where the solution is stable for several months.
- Solution 8: Ergosterol standard in KOH/methanol (ca. 200 mg l^{-1}). Dissolve ergosterol under stirring and gentle heating (50 °C) in volumetric flask, let cool, remove magnetic stirrer, adjust volume, and store at 4 °C.

3. EXPERIMENTAL PROCEDURES

3.1. Sample Preparation

1. Collect decomposing leaves, wood and/or invertebrates from streams and transport to laboratory in ice-chest.
2. Clean leaves or wood of adhering debris and macroinvertebrates; maintain animals alive in the laboratory on an appropriate diet.
3. Cut set of leaf discs with cork borer and blot lightly on filter paper; prepare representative samples of wood, shredder gut contents or faeces.
4. Place samples in extraction tubes, preserve in 10 ml KOH/methanol (Solution 2), and store in refrigerator overnight; alternatively, freeze-dry samples and analyze immediately when dry and weighed (next step is then unnecessary).

5. Prepare replicate sets of samples, dry overnight at 105 °C, and weigh to the nearest 0.1 mg in order to determine sample dry mass; optionally ash and reweigh to determine ash-free dry mass (AFDM).

3.2. Lipid Extraction and Saponification

1. Add a boiling chip to samples in KOH/methanol
2. Close tubes tightly and heat to 80 °C for 30 min.
3. Let extracts cool down (ca. 20 min).
4. To estimate recovery rates, include in each extraction series 1 leaf sample known to contain no ergosterol but spiked with 250 µl of ergosterol stock solution in KOH/methanol (Solution 8).

3.3. Conditioning of SPE Cartridges

1. Connect stop-cocks and cartridges to manifold, close stop-cocks.
2. Open pressure regulation valve of vacuum manifold.
3. Add 7.5 ml of methanol to each cartridge, start pump and open stop-cocks.
4. Increase vacuum by closing the valve, if necessary to initiate flow.
5. Close stop-cocks when about 5 mm of methanol remain above the sorbent bed.
6. Add 7.5 ml of conditioning solvent (Solution 4).
7. Open stop-cocks to suck solvent through cartridge, but leave about 5 mm above the sorbent bed.
8. Close stop-cocks and stop pump. Never let cartridge fall dry during conditioning. Restart from beginning if this happens accidentally.

3.4. Loading of Lipid Extract onto SPE Cartridge

1. Transfer lipid extract quantitatively to cartridge.
2. Rinse tube 4× with 0.5 ml methanol (e.g. syringe or Pasteur pipette).
3. Adjust volume to 12 ml with methanol if necessary.
4. Add 2 ml 0.65 M HCl (Solution 3) and make sure solutions are well mixed.
5. Open pressure regulation valve of vacuum manifold.
6. Start pump and open stop-cocks to apply vacuum.
7. Set flow rate to ≤ 1 ml min^{-1} for fastest cartridge by adjusting pressure regulation valve.
8. Control flow rate permanently and adjust if necessary.

3.5. Washing and Drying of Sorbent in SPE Cartridge

1. After complete loading of sample onto cartridge, wash sorbent bed with 2.5 ml washing solution (Solution 5).
2. Dry sorbent for 60 min under stream of air (valve completely open to achieve maximum vacuum); verify that sorbent is completely dry.

3.6. Elution of Ergosterol

1. Place HPLC vial (pre-weighed to the nearest 0.1 mg) in vacuum manifold.
2. Apply gentle vacuum and elute ergosterol with 4×400 µl of isopropanol.
3. Set flow rate to about 1 ml min^{-1} during elution.
4. Close stop-cocks, cautiously open pressure regulation valve to reduce vacuum slowly, stop pump, and remove vials form manifold.
5. Close vials tightly with corresponding cap and weigh to the nearest 0.1 mg.
6. Calculate fluid volume in vial (i.e. multiply sample weight by 1.27, given a density of isopropanol at 25 °C of 0.786 g cm^{-3}).

3.7. HPLC Analysis

1. Set chromatograph to the following conditions; retention time of ergosterol should be about 8 min:
   - Mobile phase: 100% methanol
   - Flow rate: 1.4 ml min^{-1}
   - Column temperature: 33 °C
   - Detection wavelength: 282 nm
   - Injection volume: 10 µl
2. Prepare standard curve (concentrations of 0, 5, 10, 20, and 40 µg ml^{-1}) from ergosterol stock solution in isopropanol.
3. Run standards on HPLC, then inject samples (each sample 2×).
4. Check identity of putative ergosterol peaks (1) by co-injection of the ergosterol standard with sample extract (Fig. 25.1) and (2) UV spectrometry (Fig. 25.2).
5. Measure area and/or height of ergosterol peaks.
6. Calculate ergosterol concentration in extract and leaves, based on concentration in final extract, total sample volume, and sample dry mass or ash-free dry mass.

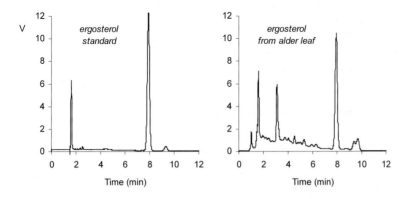

Figure 25.1. Chromatograms of an ergosterol standard and a typical lipid extract prepared from leaf litter that had been colonized by fungi in a stream.

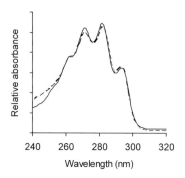

Figure 25.2. UV-absorbance spectrum of ergosterol in methanol showing the characteristic absorbance maxima of the provitamins D at about 262, 271, 282 and 294 nm. The solid line indicates a commercially available ergosterol standard; the broken line corresponds to the eluted HPLC fraction of a lipid extract prepared from leaf litter that had been colonized by fungi in a stream (from Gessner & Newell 2002).

4. FINAL REMARKS

It is important that the SPE cartridges recommended above be used with the procedure described in this chapter. Use of other cartridges is likely to result in unsatisfactory results. However, specific adaptation of the protocol (e.g. appropriate adjustment of water content of extracts) to other reversed-phase (C18) SPE cartridges with slightly different characteristics may prove feasible.

Drying of ergosterol when fixed on SPE cartridges under air (see point 3.5.) may result in substantial losses of ergosterol and inconsistent results when temperature is >21 °C. Carrying out the procedure in a cooler air-conditioned room will circumvent the problem. Replacing air by N_2 may be an alternative solution.

5. REFERENCES

Charcosset, J.-Y & Chauvet, E. (2001). Effect of culture conditions on ergosterol concentration in mycelium of aquatic hyphomycetes. *Applied and Environmental Microbiology*, 67, 2051–2055.

Gessner, M.O. & Chauvet, E. (1993). Ergosterol-to-biomass conversion factors for aquatic hyphomycetes. *Applied and Environmental Microbiology*, 85, 377–384.

Gessner M.O. & Chauvet, E. (1994). Importance of stream microfungi in controlling breakdown rates of leaf litter. *Ecology*, 75, 1807–1817.

Gessner, M.O., & Newell, S.Y. (2002). Biomass, growth rate and production of filamentous fungi in plant litter. In: C.J. Hurst, G. Knudsen, M. McInerney, L.D. Stetzenbach & M. Walter (eds.), *Manual of Environmental Microbiology* (pp. 295–308), 2nd edition. American Society for Microbiology. Washington DC.

Gessner, M.O., & Schmitt, A.L. (1996). Use of solid-phase extraction to determine ergosterol concentrations in plant tissue colonized by fungi. *Applied and Environmental Microbiology*, 62, 415–419.

Gessner, M.O., C.T. Robinson & J.V. Ward. 1998. Leaf breakdown in streams of an Alpine glacial floodplain: dynamics of fungi and nutrients. *Journal of the North American Benthological Society* 17, 403-419.

Graça, M.A.S. (2001). The role of invertebrates on leaf litter decomposition in streams – A review. *International Review of Hydrobiology*, 86, 383–393.

Hieber, M., & Gessner, M.O. (2002). Contribution of stream detritivores, fungi and bacteria to leaf breakdown based on biomass estimates. *Ecology*, 83, 1026–1038.

Maharning, A.R., & Bärlocher, F. (1996). Growth and reproduction in aquatic hyphomycetes. *Mycologia*, 88, 80–88.

Newell, S.Y. 1992. Estimating fungal biomass and productivity in decomposing litter. In: G.C. Carroll & D.T. Wicklow (eds.), *The Fungal Community. Its Organization and Role in the Ecosystem* (pp. 521–561), 2nd ed. Marcel Dekker. New York.

Suberkropp, K. (1992). Interactions with invertebrates. In: F. Bärlocher (ed.), *The Ecology of Aquatic Hyphomycetes, Ecological Studies, Vol. 94* (pp. 118–133). Springer. Berlin.

Suberkropp, K. (2001). Fungal growth, production, and sporulation during leaf decomposition in two streams. *Applied and Environmental Microbiology*, 67, 5063–5068.

CHAPTER 26

ACETATE INCORPORATION INTO ERGOSTEROL TO DETERMINE FUNGAL GROWTH RATES AND PRODUCTION

KELLER SUBERKROPP[1] & MARK O. GESSNER[2]

[1]*Department of Biological Sciences, Box 870206, University of Alabama, Tuscaloosa, AL 35487, USA;* [2]*Department of Limnology, EAWAG, Limnological Research Center, 6047 Kastanienbaum, Switzerland.*

1. INTRODUCTION

Fungi are an important component of the litter decomposer assemblages in aquatic ecosystems (Gessner et al. 1997) and important secondary producers (Suberkropp 1997, Newell 2001). Estimates of fungal growth and/or biomass production associated with decaying leaves therefore are useful to assess fungal activity. In addition, such information is essential for constructing carbon budgets for the decomposition process, and it is needed to determine organic matter turnover at the ecosystem level and to quantify the food base available to detritivorous consumers, which prefer leaves colonized by fungi.

The principal method currently available for determining growth rates and biomass production of fungi colonizing leaf litter was originally described for saltmarsh fungi by Newell & Fallon (1991) and has been modified for freshwater fungi (Suberkropp & Weyers 1996, Gessner & Chauvet 1997, Kominkova et al. 2000). An alternative method to estimate fungal production in ecosystems consists of placing leaves in the natural environment and retrieving them periodically to follow increases in fungal biomass. However, this method only gives accurate estimates if losses of fungal biomass are negligible or can be accounted for.

The method described by Newell & Fallon (1991) involves determining the rate of incorporation of radiolabelled acetate into the membrane sterol, ergosterol (Gessner & Newell 2002). Fungal growth rates are directly proportional to acetate incorporation rates and can be calculated from incorporation rates using either empirical or theoretical conversion factors (Gessner & Newell 2002). Growth rates can then be multiplied by biomass (determined from ergosterol concentrations; see

M.A.S. Graça, F. Bärlocher & M.O. Gessner (eds.), Methods to Study Litter Decomposition:
A Practical Guide, 197 – 202.

Chapter 25) to obtain fungal production. Gessner & Chauvet (1997) proposed a slightly more complicated calculation method, which is useful when growth rates are greater than about 10% per day. Since fungal biomass is generally determined as a concentration in decomposing leaf litter, production is initially calculated as fungal biomass produced per leaf dry mass or leaf ash-free dry mass per unit time. If fungal production per m^2 of stream or marsh bottom is to be determined, then the amount of leaf litter in the ecosystem must also be determined at the same time (Suberkropp 1997; see Chapter 7).

The procedure presented here has been adopted from Suberkropp & Weyers (1996; see also Suberkropp 1997). Pieces of decomposing leaves are incubated with aeration in a solution containing radiolabelled acetate. Depending on the level of activity, the incubation is ended after 2—5 hours, and the samples are placed in methanol. Ergosterol is then extracted using proper precautions for handling radioactive samples and waste. Extracted samples are injected into a high-performance liquid chromatograph (HPLC) and the concentration of ergosterol in the sample is determined. The ergosterol eluting from the HPLC is collected and the amount of radioactivity contained in the ergosterol determined with a scintillation counter.

2. EQUIPMENT, CHEMICALS AND SOLUTIONS

2.1. Equipment and Material

- Equipment and material for ergosterol extraction (see Chapter 25) plus fraction collector for HPLC
- Incubation tubes (12 mm diameter) fitted with two-holed rubber stoppers containing glass tubing for aeration. One tube for the killed control is fitted with a screw cap. Autoclave tubes before use.
- Filter apparatus with membrane filter (0.45 μm pore size) to filter-sterilize stream water
- Flow meters for each incubation tube
- Battery-operated air pump for field incubations; fishing supply stores often carry air pumps that can be operated with alkaline batteries (e.g. size D) that provide sufficient air flow for these studies.
- Set of adjustable automatic pipettes (5, 1, and 0.2 ml)

2.2. Chemicals

- Chemicals for ergosterol extraction (see Chapter 25)
- Scintillation fluid
- [1-^{14}C]acetate, sodium salt

2.3. Solutions

- Solutions for ergosterol extraction (see Chapter 25)

- Solution of [1-^{14}C]acetate plus non-radioactive sodium acetate. Adjust final concentration of acetate in the 4 ml incubation solution to 5 mM. The final specific activity should be ca. 50 MBq mmol^{-1}. Stock solution of acetate is made so that 50 μl contains ca. 1 MBq of [^{14}C]acetate and the total acetate concentration is 0.4 M.

3. EXPERIMENTAL PROCEDURES

3.1. Sample preparation and incubation with radiolabel

1. For each replicate tube, collect 3—5 decomposing leaves and cut 2 discs (ca. 12 mm diameter) from each leaf. Place one disc from each of the 3—5 leaves in a container with stream water. Avoid changes in temperature and other ambient conditions. The second leaf discs from each of the 3—5 leaves are combined, dried, weighed, and combusted to determine ash-free dry mass (see Chapters 1, 2 or 7).
2. Filter-sterilize stream water and pipette 3.95 ml into each of the incubation tubes and the control tube. Place the tubes in a rack in the stream. If handling radioactivity in the field is not possible, place tubes and leaf discs in an ice chest maintained at stream temperature and transport them back to the laboratory where they are placed in a water bath, chamber or room adjusted to stream temperature. Connect aeration tubes to a battery operated air pump and aerate each tube with 30—40 ml air min^{-1}. Alternatively, tubes can be gently shaken during the incubation.
3. To the control tube, add formalin to reach a final concentration of 2%. This tube will be used to determine the background radioactivity in the samples and is not aerated during incubation with radiolabel to avoid volatilization of formaldehyde and potential contamination of samples.
4. Add the leaf discs for each replicate to the incubation and control tubes and allow 10—20 min equilibration time.
5. Add 50 μl of [^{14}C]acetate solution to each tube at timed (e.g. 30 s) intervals. Incubate for an exact time (e.g. 180 min).
6. At timed intervals remove tubes from stream (or water bath) in the same order as above and place in an ice bath to slow further uptake of acetate. Alternatively, formalin may be added to stop acetate uptake.
7. Filter fluid and leaf discs immediately through glass fibre filters (e.g. Whatman 934-AH, 25 mm diameter). Rinse well and place filter and leaf discs in 5 ml methanol. Store at –20 °C until ergosterol is extracted.

3.2. Ergosterol extraction and determination

1. Extract ergosterol from samples following the protocol in Chapter 25, except for adjustments given below.

2. The final volume of the sample extract should be kept as small as practicable (e.g. about 500 µl). In addition, a larger volume of the extract should be injected than when only biomass is determined (e.g. 50—100 µl), and multiple (2—4) injections are recommended, especially when production rates are expected to be low. These adjustments serve to increase the sensitivity of the assay and thus estimate the radioactivity incorporated into ergosterol more precisely.

3. Collect the ergosterol peak eluting from the HPLC in a scintillation vial. This can be done either with a fraction collector that electronically detects peaks or it can be done manually. Manual collection requires precaution to avoid radioactive contamination even though radioactivity in final extracts tends to be low (typically <500 Bq).

4. Add 10 ml of scintillation fluid (e.g. Ecolume) to combined ergosterol fractions.

5. Determine radioactivity with scintillation counter and correct for quenching.

6. Calculate ergosterol concentrations from peak area and standard curve as in Chapter 25.

7. Calculate the rate of acetate incorporation into ergosterol (mmole mg^{-1} d^{-1}) as the corrected radioactivity (Bq; 1 Bq = 60 dpm; dpm = disintegrations per minute) in the sample (radioactivity of sample – radioactivity of formalin-treated control) divided by the product of the specific activity of the acetate (Bq mmole^{-1}), the fraction of the sample volume injected, time of the incubation (d), and biomass of the fungus in the sample (mg):

$$\frac{R_{sample} - R_{control}}{SA \cdot F_{sample} \cdot t \cdot B_{fungus}} \tag{26.1}$$

where: R_{sample} is the radioactivity (Bq) in the sample, $R_{control}$ is the radioactivity (Bq) in the control, SA is the specific activity (Bq/mmol), F_{sample} is the fraction of the sample that is injected into the HPLC, t is the time of the incubation in d, and B_{fungus} is the biomass of the fungus in the sample (mg).

8. Fungal growth rate (mg mg^{-1} d^{-1}) can then be calculated by multiplying the rate of acetate incorporation (as calculated above) by 19,300 mg mmol^{-1}, which is an empirically determined conversion factor (Suberkropp & Weyers 1996). See below and Gessner & Newell (2002) for a discussion of the choice of other conversion factors.

4. FINAL REMARKS

As with all procedures using radioactivity, one must use the proper precautions for purchasing, handling and disposing of radioactive materials.

Some leaf samples may have very low incorporation rates, requiring high specific activity of the radiolabelled acetate to measure incorporation reliably. As a result, assays can become exceedingly expensive. To reduce costs and radioactive waste, it is advisable to make the final ergosterol extract in as small a volume as is feasible, to inject (and collect) as large a fraction as possible, and to combine several injections for determination of radioactivity.

If formalin is used to stop acetate incorporation, samples must be filtered immediately and washed abundantly, because prolonged exposure to formalin can reduce ergosterol concentrations.

Gessner & Chauvet (1997) calculated a theoretical conversion factor of 6.6 mg mycelial biomass μmol^{-1} of incorporated acetate to convert rates of acetate incorporation into growth rates of the aquatic hyphomycete *Articulospora tetracladia*, and Newell (2000) suggested empirically determined conversion factors of 7.0—17.8 mg mycelial biomass μmol^{-1} of incorporated acetate for saltmarsh fungi (see also Gessner & Newell, 2002).

For samples from calm lentic environments, where turbulence is not needed to simulate conditions in the environment and may even curb fungal activity, aeration or shaking of incubation tubes should be avoided.

5. REFERENCES

Gessner, M.O. & Chauvet, E. (1997). Growth and production of aquatic hyphomycetes in decomposing leaf litter. *Limnology and Oceanography*, 42, 496-505.

Gessner, M.O. & Newell, S.Y. (2002). Biomass, growth rate, and production of filamentous fungi in plant litter. In C. J. Hurst, R. L. Crawford, G. Knudsen, M. McInerney, & L. D. Stetzenbach (eds.), *Manual of Environmental Microbiology* (pp. 390-408), 2nd ed. ASM Press. Washington.

Gessner, M.O., Suberkropp, K. & Chauvet, E. (1997). Decomposition of plant litter by fungi in marine and freshwater ecosystems. In D.T. Wicklow & B. Söderström (eds.), *The Mycota IV: Environmental and Microbial Relationships* (pp. 303-322). Springer-Verlag. Berlin.

Komínková, D., Kuehn, K.A., Büsing, N., Steiner, D. & Gessner, M.O. (2000). Microbial biomass, growth and respiration associated with submerged litter of *Phragmites australis* decomposing in a littoral reed stand of a large lake. *Aquatic Microbial Ecology*, 22, 271-282.

Newell, S.Y. (2000). Methods for determining biomass and productivity of mycelial marine fungi. In K.D. Hyde, & S.B. Pointing (eds.), *Marine Mycology – A Practical Approach* (pp. 68-90). Fungal Diversity Press. Hong Kong.

Newell, S.Y. (2001). Multiyear patterns of fungal biomass dynamics and productivity within naturally decaying smooth cordgrass shoots. *Limnology and Oceanography*, 46, 573-583.

Newell, S.Y., & Fallon, R.D. (1991). Toward a method for measuring fungal instantaneous growth rates in field samples. *Ecology*, 72, 1547-1559.

Suberkropp, K. (1997). Annual production of leaf-decaying fungi in a woodland stream. *Freshwater Biology*, 38, 169-178.

Suberkropp, K. & Weyers, H. (1996). Application of fungal and bacterial production methodologies to decomposing leaves in streams. *Applied and Environmental Microbiology*, 62, 1610-1615.

CHAPTER 27

BACTERIAL COUNTS AND BIOMASS DETERMINATION BY EPIFLUORESCENCE MICROSCOPY

NANNA BUESING

Department of Limnology, EAWAG, Limnological Research Center, 6047 Kastanienbaum, Switzerland.

1. INTRODUCTION

Accurate estimates of bacterial abundance and biomass are a critical prerequisite for assessing the roles of bacteria in biogeochemical cycles and food webs. In addition, they are important for understanding bacterial population dynamics in natural systems, including litter decomposition systems in streams, wetlands and other environments. The most widely used approach to obtain such estimates is to pass a bacterial suspension through a membrane filter, stain the trapped cells with a fluorescent dye and count them under an epifluorescence microscope. When bacteria are associated with particles such as sediments and decomposing litter, it is best to first detach cells quantitatively from their substrate before counting them (Fry 1988), which in most cases is efficiently achieved with ultrasonic probes (e.g. Velji & Albright 1986, Buesing & Gessner 2002).

Specific and intense staining of bacteria is required to facilitate clear differentiation between bacterial cells and other particles. This is especially critical when samples are rich in organic matter. Traditionally, acridine orange (AO) and 4',6-diamidino-2-phenylindole (DAPI) have been used for that purpose. However, in recent years, a variety of more suitable RNA and DNA binding dyes have become commercially available. These include SYBR Green I and II, YOYO-1, YO-PRO-1, SYTO, and PicoGreen. The specificity and staining intensity of these dyes are much greater than those of DAPI and AO, facilitating recognition and quantification of bacteria significantly.

The detached and stained bacterial cells are viewed with an epifluorescence microscope, counted, sized and assigned to size classes. Alternatively, digital images can be taken and analyzed with an image analysis system. In addition to reducing observer bias, the image analysis approach offers the advantage that cell dimensions

M.A.S. Graça, F. Bärlocher & M.O. Gessner (eds.), Methods to Study Litter Decomposition: A Practical Guide, 203 – 208.

and shapes needed to determine biomass are measured for each individual bacterium, circumventing the need to delineate a limited number of size classes.

The detachment procedure described here is adopted from Buesing & Gessner (2002). The staining procedures follow protocols developed for counting viruses in water samples by Noble & Fuhrman (1998), Weinbauer et al. (1998) and Lebaron et al. (1998), and applied to decomposing litter by Buesing (2002).

2. EQUIPMENT, CHEMICALS AND SOLUTIONS

2.1. Equipment and Material

- Epifluorescence microscope equipped with a high-pressure mercury lamp (HPO 50 W or, preferably, HPO 100 W)
- Optical filter set for detection of stained cells (e.g. Chroma light filter set no. 41001; excitation filter 480 nm, beam splitter 505 nm, emission filter 530 nm)
- Image analysis system: cooled charge coupled device (CCD) camera (e.g. Photometrics SenSys® 0400, Roper Scientific, Trenton, USA), frame grabber, computer and image analysis software (e.g. MetaMorph Imaging Software, Universal Imaging Corp., Downingtown, USA)
- Aluminium oxide membrane filters (Whatman Anodisc, 0.2 µm pore size, 25 mm diameter)
- Membrane filters of cellulose nitrate or mixed cellulose esters (e.g. Millipore HAWP, 0.45 µm pore size, 25 mm diameter)
- Forceps for handling filters
- Clean boxes to collect leaf material
- Sterile 20 ml glass vials for leaf discs or litter pieces
- Sterile Eppendorf tubes
- Micropipettes and sterile tips for dispensing volumes of 10—1000 µl
- Ultrasonic probe (e.g. Branson Sonifier 250)
- Vortex
- Filter manifold with straight filtration funnels
- Vacuum pump
- Slides and cover slips
- Petri dishes (60 mm diameter)
- Small cardboard box

2.2. Chemicals

- SYBR Green II (Molecular Probes, Eugene, Oregon, USA)
- Ethanol, technical grade
- Formaldehyde (37%), analytical grade
- Non-fluorescing immersion oil (e.g. Zeiss, Immersol 518N)
- Sodium pyrophosphate ($Na_4P_2O_7$), analytical grade

- Glycerol, analytical grade
- NaCl, analytical grade
- NaH_2PO_4, analytical grade
- *p*-Phenylenediamine ($C_6H_8N_2$)

2.3. Solutions

- Particle-free water (e.g. 0.2 µm filtered autoclaved Nanopure® water)
- Staining solution 1 (original SYBR Green II solution diluted 1:10 in Nanopure water)
- Staining solution 2 (2.5% working solution from staining solution 1); prepared the day when samples are stained
- Fixation solution (2% formalin, 0.1% sodium pyrophosphate)
- Antifade mounting solution: 50% glycerol, 50% phosphate-buffered saline (PBS; 120 mM NaCl, 10 mM NaH_2PO_4, pH 7.5), 0.1% *p*-phenylenediamine

3. EXPERIMENTAL PROCEDURES

3.1. Sample Preparation

1. Place small litter pieces such as leaf discs (~100—500 mg wet mass; corresponding to about 20—100 mg dry mass) into glass scintillation vials, add 10 ml of fixation solution and store at 4 °C for a maximum of 3 months, preferably shorter.
2. Take a second set of identical subsamples for determining relationships between sample wet mass and dry mass or between surface area and dry mass.
3. Place sample on ice and sonicate for 1 min with an ultrasonic probe (output: 80 W and 76 µm amplitude).
4. Clean ultrasonic probe carefully with ethanol before treating the next sample.
5. Mount cellulose filter onto the filtration manifold and rinse with a small volume of Nanopure water.
6. Place the Anodisc filter flat on top of the moist cellulose filter.
7. Connect filtration funnel and add 1 ml Nanopure water.
8. Vortex sample, wait for 10 s, take a 10—400 µl aliquot from 2 mm below the surface, and add the aliquot to the Nanopure water in the filtration funnel.
9. Add another 1 ml of Nanopure water to ensure good mixing of the sample suspension prior to filtration.
10. Filter sample through the Anodisc filter by applying a vacuum of max. 20 kPa (200 mbar).
11. Dry filter carefully by placing it on a cleansing tissue.

3.2. Staining

1. Pipette 100 μl of staining solution 2 in a clean Petri dish and place Anodisc filter on top.
2. Cover Petri dish with the cardboard box to keep the sample in the dark during staining.
3. Stain for 15 min.
4. Dry filter again by placing it on a cleansing tissue.
5. Mount filter on a clean slide, add 30 μl of antifading solution and place a cover slip on top.
6. Press gently on the cover slip with forceps until the antifading solution is evenly distributed.

3.3. Counting and Cell Sizing by Image Analysis

Cell numbers should be determined in 10—20 microscopic fields (typically >400 cells; Kirchman 1993). From each microsopic field, a digital image is taken with a cooled charge coupled device (CCD) camera. Digital images are electronically stored and analyzed with image analysis software.

First, captured digital images are contrast-stretched, which means that a range of gray values of the 16-bit image is selected and scaled to 256 gray levels, thereby increasing the apparent contrast in the resulting image. Next, cell edges are sharpened by applying a high-pass filter (e.g. a "Mexican hat" kernel). A similar effect is produced when a low-pass filter is used and the resulting image is subsequently subtracted from the original picture. The optimal threshold defining the objects is set manually. The resulting binary picture can then be edited by erode and dilate functions, which will separate touching cells or fill small holes, respectively. The final editing is interactive in an overlay mode with the originally captured image.

Once the edited binary image is completed, the area and perimeter of each cell is determined to derive cell lengths (l) and widths (w). This indirect procedure to determine l and w is necessary, because imaging software systems generally overestimate real cell dimensions severely, especially when bacteria are curved (Massana et al. 1997). Cell volumes (V) of individual cells may be calculated under the assumption that cells are cylinders with hemispherical ends (Fry 1988), which works for both rods and cocci.

3.4. Manual Counting and Sizing

If bacterial cells are counted manually, then size classes according to size and shape need to be defined. Bacteria in each optical field are counted in each of these established size classes. Again at least 10—20 microscopic fields (typically >400 cells; Kirchman 1993) should be counted.

3.5. Calculations

The total biovolume (BV) of bacterial cells per g of leaf material is calculated as follows:

$$\frac{BV}{DM_l} = \frac{(\sum bv_i) \cdot V_s \cdot A_f}{S_f \cdot A_c \cdot DM_l} \tag{27.1}$$

where bv_i is the biovolume of individual bacterial cell, V_s the sample volume, A_f the total filtration area, S_f the volume of the subsample passed over the filter, A_c the filtration area, in which bacteria were counted, and DM_l the litter dry mass.

Bacterial dry mass or bacterial carbon is calculated from bacterial BV based on empirically determined conversion factors. For pelagic freshwater bacteria, Loferer-Krößbacher et al. (1998) established the following relationship:

$$dm_b = 435 \cdot bv^{0.86} \tag{27.2}$$

where dm_b is the dry mass and bv the biovolume of a bacterial cell. In the size range of cells expected for bacteria associated with organic matter, this conversion factor results in slightly higher estimates of bacterial biomass than most other published conversion factors.

4. FINAL REMARKS

Control counts of Nanopure water without sample must be run daily and for every new batch of filters and stain to check for possible contamination. Controls without samples should also be prepared every time samples are preserved for later counts.

When counting stained cells, care must be taken to ensure that all filtered bacteria appear in a single optical layer. Since Anodisc filters have a distinctive ring around the actual filtration area; there is a risk of producing multiple layers when adding the antifading solution. This is best avoided by adding the antifading solution after the filter has completely dried.

A number of factors for converting bacterial biovolume to biomass have been published (e.g. Fagerbakke et al. 1996, Theil-Nielsen & Søndergaard 1998, Vrede et al. 2002). Most of these have been derived from either *E. coli* in culture, which favours large cells, or from bacteria sampled in marine pelagic environments, where cells tend to be small. A specific conversion factor for bacteria associated with litter is not available. Some general aspects of choosing factors for converting bacterial biovolume to biomass are discussed by Norland (1993).

5. REFERENCES

Buesing (2002). Microbial productivity and organic matter flow in a littoral reed stand. Dissertation ETH Zurich No. 14667.

Buesing, N. & Gessner, M.O. (2002). Comparison of detachment procedures for direct counts of bacteria associated with sediment particles, plant litter and epiphytic biofilms. *Aquatic Microbial Ecology*, 27, 29-36.

Fagerbakke, K.M., Heldal, M. & Norland, S. (1996). Content of carbon, nitrogen, oxygen, sulfur and phosphorus in native aquatic and cultured bacteria. *Aquatic Microbial Ecology*, 10, 15-27.

Fry, J.C. (1988) Determination of biomass. In: B. Austin (ed.) *Methods in Aquatic Bacteriology* (pp. 27-72). John Wiley & Sons. New York.

Kirchman, D.L. (1993). Statistical analysis of direct counts of microbial abundance. In: P.F. Kemp, B.F. Sherr, E.B. Sherr, & J.J. Cole (eds.). *Handbook of Methods in Aquatic Microbial Ecology* (pp. 117-119). Lewis Publishers. Boca Raton.

Lebaron, P., Parthuisot, N. & Catala, P. (1998). Comparison of blue nucleic acid dyes for flow cytometric enumeration of bacteria in aquatic systems. *Applied and Environmental Microbiology*, 64, 1725-1730.

Loferer-Krößbacher, M., Klima, J. & Psenner, R. (1998). Determination of bacterial cell dry mass by transmission electron microscopy and densiometric image analysis. *Applied and Environmental Microbiology*, 64, 688-694.

Massana, R., Gasol, J.M., Bjørnsen, P.K., Blackburn, N., Hagström, Å., Hietanen, S., Hygum, B.H., Kuparinen, J. & Pedrós-Alió, C. (1997). Measurement of bacterial size via image analysis of epifluorescence preparations: description of an inexpensive system and solutions to some of the most common problems. *Scientia Marina*, 61, 397-407.

Noble, R.T. & Fuhrman, J.A. (1998). Use of SYBR Green I for rapid epifluorescence counts of marine viruses and bacteria. *Aquatic Microbial Ecology*, 14, 113-118.

Norland, S. (1993). The relationship between biomass and volume of bacteria. In: P.F. Kemp, B.F. Sherr, E.B. Sherr, J.J. Cole (eds.). *Handbook of Methods in Aquatic Microbial Ecology* (pp. 303-307). Lewis Publishers. Boca Raton.

Theil-Nielsen, J. & Søndergaard, M. (1998). Bacterial carbon biomass calculated from biovolumes. *Archiv für Hydrobiologie*, 2, 195-207.

Velji, M.I. & Albright, L.J. (1986). Microscopic enumeration of attached marine bacteria of seawater, marine sediment, fecal matter, and kelp blade samples following pyrophosphate and ultrasound treatments. *Canadian Journal of Microbiology*, 32, 121-126.

Vrede, K., Heldal, M., Norland, S. & Bratbak, G. (2002). Elemental composition (C, N, P) and cell volume of exponentially growing and nutrient-limited bacterioplankton. *Applied and Environmental Microbiology*, 68, 2965-2971.

Weinbauer, M.G., Beckmann, C. & Höfle, M.G. (1998). Utility of green fluorescent nucleic acid dyes and aluminum oxide membrane filters for rapid epifluorescence enumeration of soil and sediment bacteria. *Applied and Environmental Microbiology*, 64, 5000-5003.

CHAPTER 28

SECONDARY PRODUCTION AND GROWTH OF LITTER-ASSOCIATED BACTERIA

NANNA BUESING & MARK O. GESSNER

Department of Limnology, EAWAG, Limnological Research Center, 6047 Kastanienbaum, Switzerland.

1. INTRODUCTION

Bacterial secondary production (BSP) can constitute a large fraction of total secondary production associated with decomposing plant litter (Buesing 2002), suggesting that BSP can be a crucial component of total carbon flow in systems where plant litter is an important source of organic matter. Quantifying bacterial secondary production and/or growth is therefore important for addressing many ecological questions, including the assessment of the role and dynamics of bacteria in decomposing litter systems. In conjunction with estimates of bacterial biomass, BSP allows calculating bacterial growth rates, a key parameter to describe bacterial population dynamics.

Various methods have been used to estimate bacterial production. The two most common ones are the thymidine method (Fuhrman & Azam 1980) and the leucine method (Kirchman et al. 1985). Both are based on measuring the incorporation of radiolabelled precursor molecules into macromolecules of the bacterial cell over a known period. [Methyl-^{3}H]thymidine is used to determine rates of DNA synthesis. [^{3}H]leucine or [^{14}C]leucine and sometimes other radiolabelled amino acids are used to estimate rates of protein synthesis. The underlying assumption of both approaches is that the synthesis rate of macromolecules is directly proportional to cell growth. Riemann & Bell (1990), Robarts & Zohary (1993), Chin-Leo (2002) and others discuss general theoretical and practical aspects of these methods.

The present chapter presents a procedure that is applicable to bacteria associated with particulate organic matter, such as plant litter derived from macrophyte tissues or leaves from riparian trees (Buesing & Gessner 2003). The method is based on the incorporation of radiolabelled leucine into protein. It has several advantages. In

M.A.S. Graça, F. Bärlocher & M.O. Gessner (eds.), Methods to Study Litter Decomposition: A Practical Guide, 209 – 216.

particular, protein represents a very large and rather constant fraction of bacterial carbon (Simon & Azam 1989). As a result, conversion factors for calculating bacterial biomass production from leucine incorporation rates vary only twofold or less (Simon & Azam 1989, Moran & Hodson 1992) compared to a tenfold variation in conversion factors for thymidine incorporation rates (Riemann & Bell 1990). The main potential shortcoming of the leucine method is that its specificity resides only in the greater capacity of bacteria to take up organic molecules efficiently at much lower concentrations than eukaryotic organisms.

In practice, radioactive leucine is added to the litter sample and leucine is incorporated into bacterial protein during a short incubation period. The incorporation is stopped by adding trichloroacetic acid (TCA). Samples are sonicated and the liquid is removed from the litter sample and passed through a membrane filter. Both the filter and plant litter are then successively washed with TCA, a non-radioactive leucine solution, ethanol and water. Finally, the filter and litter are combined, and the protein is dissolved in hot alkaline solution and radioassayed. The specific protocol presented here is based on the method described in Buesing & Gessner (2003), adapted for plant litter samples.

2. EQUIPMENT, CHEMICALS AND SOLUTIONS

2.1. Equipment and Material

- Polycarbonate filter (e.g. Millipore GTTP, 0.2 μm pore size, 25 mm diameter)
- Cellulose membrane filters filters (e.g. Millipore HAWP 0.45 μm pore size, 25 mm diameter)
- Cork borer, cutter or scissors for subsampling plant litter
- Sterile 20 ml glass vials
- Sterile glass bottle for storing lake or stream water
- Analytical balance (preferably 0.01 mg precision)
- Aluminium or porcelain dishes for determining sample fresh mass, dry mass and ash mass
- Drying oven (105 °C)
- Micropipettes and sterile tips for dispensing volumes between 10 and 1000 μl
- 2-ml screw-cap microcentrifuge tubes
- Ultrasonic probe (e.g. Branson Sonifier 250, output 80 W, amplitude 76 μm)
- Filter manifold
- Dry block heater (90 °C)
- Benchtop centrifuge (14000 g)
- Plastic scintillation vials (20 ml)
- Scintillation counter

2.2. Chemicals

- [^{3}H]leucine (sterile; specific activity 4.4—7 TBq mmol^{-1}) or [^{14}C]leucine (specific activity >11 GBq mmol^{-1})
- L-leucine, analytical grade
- Trichloroacetic acid (TCA)
- Ethanol, analytical grade
- NaOH, analytical grade
- Ethylenediamine tetra acetic acid (EDTA), analytical grade
- Sodium dodecyl sulfate (SDS), analytical grade
- Scintillation cocktail for counting samples containing TCA and NaOH (e.g. HionicFluor™, Packard Bioscience, Meriden, USA)

2.3. Solutions

- 1.5 mM leucine (4.5 µM radioactive leucine plus non-radioactive leucine, specific activity 6—8 · 10^9 Bq mmol^{-1}) for 3 ml incubation volume per sample
- 40 mM L-leucine (non-radioactive)
- 50% TCA
- 5% TCA
- 80% ethanol
- Deionized water (e.g. Nanopure® water)
- Alkaline extraction solution: 0.5 M NaOH, 25 mM EDTA, 0.1% SDS

3. EXPERIMENTAL PROCEDURES

3.1. Sample Preparation

1. Collect litter samples in the field and keep them at *in situ* temperature during transport, processing and incubation with radiolabel; avoid also other changes in environmental conditions and process samples as quickly as possible, preferably in the field.
2. Take representative subsample from plant material; e.g., cut discs from leaf litter with a cork borer or cut bulk litter samples into small pieces with a cutter or scissors.
3. Take three sets of subsamples, one for measuring leucine incorporation rates into bacterial production, one for a control in which microbes have been killed with TCA before the addition of leucine, and one for establishing area-dry mass relationships or fresh mass-dry mass relationships.
4. Place two sets of subsamples each with 5 leaf discs or ~75 mg litter wet mass (corresponding to 15 mg dry mass) into a 20-ml glass scintillation vial.
5. Add 2.9 ml of filtered (0.2 µm pore-size membrane filter) stream or lake water.
6. Place subsamples for determining dry mass in aluminium or porcelain dishes and dry at 105 °C to constant weight.

3.2. Incubation

1. Add TCA to controls (final concentration of 5%) before incubating samples.
2. Add 0.1 ml of 1.5 mM leucine (mixture of radioactive and non-radioactive leucine at a final concentration of 50 μM) to each sample at timed intervals (e.g. 30 s).
3. Incubate samples for 30 min at *in situ* temperature.
4. Stop leucine incorporation at timed intervals as above by the addition of TCA to a final concentration of 5%.

3.3. Purification and Extraction

1. Place samples on ice.
2. Sonicate samples for 1 min.
3. Place samples back on ice for at least 15 min.
4. Transfer the 3-ml sample to the filtration manifold and filter onto a 0.2-μm polycarbonate filter backed by a cellulose filter.
5. Wash plant litter in the glass vial with 1 ml of 5% TCA.
6. Transfer washing volume to the filtration manifold and filter through the same polycarbonate filter.
7. Repeat TCA washing step.
8. Wash sample with 1 ml of 40 mM non-radioactive leucine solution, 1 ml of ethanol and 1 ml of Nanopure water, each time transferring the washing volume onto the filter, and apply vacuum.
9. Transfer plant material and polycarbonate filter to a 2-ml screw-cap microcentrifuge tube.
10. Add 1 ml of the alkaline extraction solution.
11. Heat samples for 60 min at 90 °C in a dry block heater to dissolve proteins.
12. Cool samples down to ambient temperature.
13. Pipette a 100—500 μl aliquot into a scintillation vial.
14. Add 5 ml of HionicFluor™ scintillation cocktail and determine radioactivity in the scintillation counter.

3.4. Calculation

Calculate leucine incorporated into bacterial protein (leu$_{inc}$) in mol per g litter dry mass per day as follows:

$$leu_{inc} = \frac{\left(dpm_{sample} - dpm_{control}\right) \cdot V}{SA \cdot t \cdot v \cdot DM} \tag{28.1}$$

where *dpm* = disintegrations per minute converted from measured *cpm* = counts per minute (see remarks below; 1 Bq = 60 dpm), *V* = total volume of the extract (ml), *SA* = specific activity of the final leucine solution (dpm mol^{-1}), *t* = incubation time (hour), *v* = aliquot of counted sample (ml), *DM* = dry mass of plant material (g).

Bacterial secondary production (BSP) in g C per g litter dry mass per day is then calculated as follows:

$$BSP = leu_{inc} \cdot \frac{b}{a} \cdot c \cdot ID \qquad (28.2)$$

where a = molar fraction of leucine in protein (0.073) (Simon & Azam 1989), b = the molecular weight of leucine (131.2 g mol^{-1}), c = weight fraction of cellular carbon in protein (0.86; Simon & Azam 1989), and ID = isotope dilution (e.g. 1.23 for plant litter; Buesing & Gessner 2003).

4. FINAL REMARKS

Preliminary tests must be run to establish: (a) whether leucine is incorporated into protein at a constant rate; this is tested by incubating samples with an appropriate concentration of total leucine for different time periods (e.g. 5—90 min); incorporation of radiolabel should increase linearly with time. (b) That incubations are carried out at total leucine concentrations that maximize leucine incorporation rates; this is done by incubating samples with different total leucine concentrations while keeping the molar ratio of radioactive and non-radioactive leucine constant. Saturation curves are obtained by plotting incorporation rates vs. the total leucine concentration. In litter and sediment samples, saturation may occur at concentrations as high as 50 µM (Marxsen 1996, Fischer & Push 1999, Buesing & Gessner 2003), although saturation at a lower concentration (400 nM) has been found with leaves decomposing in a stream (Suberkropp & Weyers 1996).

Isotope dilution is the dilution of the added radiotracer by the extracellular and/or intracellular pools of that substance. It may be determined either by linearly regressing the reciprocal of incorporated radioactivity against leucine concentration (Bird 1999) or by nonlinear regression analysis of leucine saturation curves (Van Looij & Riemann 1993, Buesing & Gessner 2003).

A quench curve for each type of sample can be established by extracting non-radioactive organic matter using the same procedure as described above. Constant amounts of radioactivity are then added to an increasing volume of a sample extract. Total volumes are kept constant by adding to the sample extract appropriate volumes of NaOH. Samples are then radioassayed. Subsequently, the 'transformed spectral index of the external standard' (tSIE, Packard scintillation counter) is plotted against the counting efficiency, and the resulting quench curve is used to convert cpm to dpm.

Tritiated leucine has been used in most applications to date, probably because it is much cheaper than [^{14}C]leucine. However, the use of [^{14}C]leucine is preferable from a theoretical point of view, because the decay energy of ^{14}C is much higher (beta maximum energy of 156 keV) than that of ^{3}H (18.6 keV), resulting in a higher counting efficiency (~90% compared to ~60%). However, if only few bacterial cells are active, leucine with a high specific activity may have to be used. In this case, the use of [^{3}H]leucine may be advantageous, because it can be purchased at a specific activity (4.4—7 TBq mmol^{-1}) nearly 1000 times higher than [^{14}C]leucine with only ~11 GBq mmol^{-1}.

It is possible to reduce the incubation volume from 3 to 1 ml in order to reduce the required amount of radiolabelled leucine and hence costs.

5. REFERENCES

Bird, D.F. (1999). A critical examination of substoichiometric isotope dilution analysis using thymidine and leucine. *Scientia Marina*, 63, 61-70.

Buesing, N. (2002). Microbial productivity and organic matter flow in a littoral reed stand. Dissertation ETH Zurich No. 14667.

Buesing, N. & Gessner, M.O. (2003). Incorporation of radiolabeled leucine into protein to estimate bacterial production in plant litter sediment epiphytic biofilms and water samples. *Microbial Ecology*, 45, 291-301.

Chin-Leo, G. (2002). Bacterial secondary productivity. In: C.J. Hurst, R.L. Crawford, G.R. Knudsen, M.J. McInerney & L.D. Stetzenbach (eds.) *Manual of Environmental Microbiology*, 2nd ed. (pp. 354-363). ASM Press. Washington, DC.

Fischer, H. & Pusch, M. (1999). Use of the [^{14}C]leucine incorporation technique to measure bacterial production in river sediments and the epiphyton. *Applied and Environmental Microbiology*, 65, 4411-4418.

Fuhrman, J.A. & Azam, F. (1980). Bacterioplankton secondary production estimates for coastal waters of British Columbia, Antarctica, and California. *Applied and Environmental Microbiology*, 39, 1085-1095.

Kirchman, D.L., K'Nees, E. & Hodson, R. (1985). Leucine incorporation and its potential as a measure of protein synthesis by bacteria in natural aquatic systems. *Applied and Environmental Microbiology*, 49, 599-607.

Marxsen, J. (1996). Measurement of bacterial production in stream-bed sediments via leucine incorporation. *FEMS Microbiology Ecology*, 21, 313-325.

Moran, M.A. & Hodson, R.E. (1992). Contributions of three subsystems of a freshwater marsh to total bacterial secondary productivity. *Microbial Ecology*, 24, 161-170.

Riemann, B. & Bell, R.T. (1990). Advances in estimating bacterial biomass and growth in aquatic systems. *Archiv für Hydrobiologie*, 118, 385-402.

Robarts, R.D. & Zohary, T. (1993). Fact or fiction – bacterial growth rates and production as determined by [methyl-^{3}H]thymidine? *Advances in Microbial Ecology*, 13, 371-425.

Simon, M. & Azam, F. (1989). Protein content and protein synthesis rates of planktonic marine bacteria. *Marine Ecology Progress Series*, 51, 201-213.

Suberkropp, K. & Weyers, H. (1996). Application of fungal and bacterial production methodologies to decomposing leaves in streams. *Applied and Environmental Microbiology*, 62, 1610-1615.

van Looij, Q. & Riemann, B. (1993). Measurements of bacterial production in coastal marine environments using leucine: application of a kinetic approach to correct for isotope dilution. *Marine Ecology Progress Series*, 102, 97-104.

CHAPTER 29

ISOLATION OF CELLULOSE-DEGRADING BACTERIA

JÜRGEN MARXSEN

Limnologische Fluss-Station des Max-Planck-Instituts für Limnologie, D-36110 Schlitz, Germany.

1. INTRODUCTION

The details of many microbial processes can only be clarified by using pure cultures. Such processes include symbiotic activities during the decomposition of cellulose and other polymeric plant constituents. Hence, to assess the specific roles of microbial species in plant litter decomposition, it is important that these organisms are available as pure cultures.

The term *pure culture* means that all its constituent cells are descendants of the same individual. A pure culture is therefore genetically pure (although mutations can lead to genetic changes during storage, especially in growing cultures). *Axenic cultures*, in contrast, contain cells of a single species, free of any other living organisms but not necessarily consisting of genetically identical individuals (Pelczar & Chan 1977).

A pure culture can be obtained by using a micromanipulator in combination with a microscope, but in most cases indirect methods are applied (Rodina 1972, Schneider & Rheinheimer 1988, Overmann 2003). Samples are commonly inoculated on selective media that allow the target organism to multiply while inhibiting or preventing most other organisms. A pure culture results from this approach only, however, if the microbial population of the colony of interest has grown from a single cell. This may not always be the case. Thus, it is necessary to further examine the culture by microscopic, cultural or biochemical tests to ascertain its genetic purity (Trüper et al. 2001).

The basic technique to obtain a pure bacterial culture is illustrated in this chapter for cellulose-degrading bacteria, since cellulose is the most common plant polymer in nature. Cellulose is rather resistant to biological attack and is only degraded by a small subset of bacteria. Numerous descriptions for cultivating and isolating cellulose-degrading bacteria have been published (Reichenbach & Dworkin 1981,

M.A.S. Graça, F. Bärlocher & M.O. Gessner (eds.), Methods to Study Litter Decomposition: A Practical Guide, 217 – 222.
© 2005 *Springer. Printed in The Netherlands.*

Reichenbach 1999). The method selected here has been adopted from Reichenbach & Dworkin (1981) and Schneider & Rheinheimer (1988).

2. EQUIPMENT, CHEMICALS, SOLUTIONS AND MEDIA

2.1. Equipment

- Safety-cabinet or laminar flow hood
- Autoclave
- Incubator (preferably with cooling system)
- Drying oven for sterilization of glassware (160—180 °C)
- Sterile Petri dishes (90—100 mm diameter)
- Glass spreaders (Drigalski spatula)
- Inoculation loops
- Test tubes
- Vortex mixer
- Erlenmeyer flasks
- Pipettes (1 and 10 ml)

2.2. Chemicals (reagent grade or better)

- Deionized water
- Agar
- Mineral salts (see section 2.3.)
- Ethylenediaminetetraacetic acid (EDTA), iron(III) sodium salt (trihydrate)
- Cycloheximide
- KOH
- HCl
- Powdered cellulose (e.g. MN 300 from Macherey & Nagel, Düren, Germany)
- Filtered water from sampling site

2.3. Solutions and Media

- Trace element solution (Reichenbach & Dworkin 1981, Drews 1974): 100 mg $MnCl_2 \cdot 4H_2O$, 20 mg $CoCl_2$, 10 mg $CuSO_4$, 10 mg $Na_2MoO_4 \cdot 2H_2O$, 20 mg $ZnCl_2$, 5 mg LiCl, 5 mg $SnCl_2 \cdot 2H_2O$, 10 mg H_3BO_3, 20 mg KBr, 20 mg KI, 8 g EDTA iron(III) sodium salt (trihydrate), distilled water to 1 l. Sterilize by filtration.
- Stanier mineral agar (Stanier 1942, Reichenbach & Dworkin 1981): 1.0 g $(NH_4)_2SO_4$, 1.0 g K_2HPO_4 (autoclave separately), 0.2 g $MgSO_4 \cdot 7H_2O$, 0.1 g $CaCl_2 \cdot 2H_2O$, 0.02 g $FeCl_3$, 1 ml trace element solution, distilled water to 1 l. After sterilization adjust pH to 7.0—7.5 with 1 M NaOH or HCl, if necessary. In order to suppress the growth of fungi 25 mg l^{-1} cycloheximide (filter-sterilized) can be added after autoclaving (Brockman 1967).

- Cellulose overlay agar: 0.4% powdered cellulose in Stanier mineral agar, poured as thin layer on top of Stanier mineral agar without cellulose.

3. EXPERIMENTAL PROCEDURES

3.1. Preparation of Media and Incubation of Agar Plates

1. Pour the mineral medium without cellulose into sterile Petri dishes. After it has gelled, pour a thin layer of cellulose overlay agar on top.
2. Separate bacterial cells from the plant litter. This can be done by cutting the plant material into small pieces and shaking them in sterile water from the sampling site containing 0.1% sodium pyrophosphate or detergents (e.g. Tween). Additionally, samples can be treated with an ultrasonic probe, a tissue homogenizer (e.g. Ultra-Turrax or Polytron), or a laboratory blender (Buesing & Gessner 2002, Chapter 27). Special care needs to be taken during the detachment step to apply stringent aseptic procedures to ensure that native bacteria present in the plant litter sample are recovered, rather than contaminants.
3. Depending on the cell concentration, it may be necessary to dilute the bacterial suspension resulting from the detachment procedure. To this end, prepare at least five test tubes each with 9 ml of autoclaved water from the sampling site. With a sterile pipette, transfer 1 ml of the original cell suspension to the first dilution tube. Vortex and transfer 1 ml of the diluted suspension to a new tube containing 9 ml of sterile water. Repeat this procedure several times to obtain a dilution series.
4. With a flamed loop, place a drop of a suspension (i.e. the inoculum) in the centre of an agar plate. Alternatively, transfer the inoculum (e.g. 0.5 ml) to the medium with a sterile pipette. In both cases, spread the drop over the entire agar surface of the dish with a sterile Drigalski spatula. Use suspensions from different dilution steps for different plates.
5. Incubate the Petri dishes at temperatures between 12 and 30 °C (or lower), depending on the origin of the samples. Incubation at *in situ* temperature is generally preferable, as it favours the development of typical strains active in the natural environment. A drawback of incubation at low temperature is that colonies develop much more slowly than colonies incubated at room or higher temperature.
6. Check the plates at regular intervals. The first colonies may appear within 24 h, or after several weeks, or even months at low incubation temperature.
7. Especially when attempting to isolate cellulose-degrading bacteria, many gliding organisms (mainly from the *Cytophaga* group) typically occur. They may spread over the agar surface and are often difficult to recognize because they are thin, translucent, lack sharp borders, and are mostly yellowish to whitish. To prevent mixing of colonies, check plates regularly. If mixing is detected, the colonies of interest should be transferred to fresh medium.

3.2. Isolation

1. Use a flamed loop to transfer bacterial cells from a clearly separated colony to a new plate and spread the inoculum over an area A of the plate as shown in Fig. 29.1. After flaming the loop, make a streak through a small section of A to spread some of the inoculum to area B. Flame the loop again and streak through a section of B to further spread the inoculum over C. Each time the loop is flamed, let it cool down before using it again, to avoid killing cells by heat.
2. Repeat this step several times by streaking material from newly grown, clearly separated colonies on a fresh plate, until all colonies on the final plate look identical.

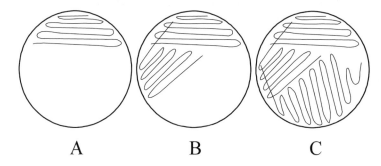

A   B   C

Figure 29.1: Example scheme for streaking inocula on agar surfaces. A, B and C = first and subsequent steps.

3. A purity test of the culture is essential. Isolates are commonly regarded as pure if (i) the same uniform colony type develops in subsequent subcultures, and if (ii) the cells are morphologically uniform (Schneider & Rheinheimer 1988). Advice by an experienced microbiologist is recommended. A range of molecular techniques (e.g. Trüper et al. 2001) can be used to enhance confidence in the purity of cultures.
4. After a pure bacterial culture has been obtained, it is normally necessary to maintain it in a viable state for an extended period (weeks to years). Short-term preservation during ongoing work is possible by periodic transfer to fresh medium. Lyophilization and subsequent storage under vacuum (Schneider & Rheinheimer 1988) or storage in liquid nitrogen (Hespell & Bryant 1981, Pfennig & Trüper 1989) are useful techniques for long-term preservation (years to decades).

4. FINAL REMARKS

Cellulose agar overlays are used to prevent the cellulose from settling on the bottom of the dish where it is out of reach for the developing colonies. Cell–fiber contact is essential for cellulose attack by bacteria (Reichenbach & Dworkin 1981).

It is crucial to maintain stringent aseptic conditions during all isolation procedures. Otherwise there is a large risk of cultivating contaminants rather than the wanted native bacteria.

Instead of detaching cells from plant litter, it is possible to place small (e.g. 2—3 mm diameter) plant pieces directly on cellulose agar plates. A mixture of bacterial strains develops around such a piece. From this mixture a small amount can be transferred to a new agar plate where the cells are streaked out as described in 3.2.

By using other growth media, many different types of microorganisms can be isolated. Such media and alternative procedures are readily found in the microbiological literature (e.g. Rodina 1972, Starr et al. 1981, Austin 1988, Dworkin et al. 1999-2003). It is important to realize, however, that most bacteria still resist isolation and culturing by current techniques (Head et al. 1998). This remains true despite the development of several sophisticated isolation techniques in recent years (Overmann 2003).

5. REFERENCES

Austin, B. (ed.) (1988). *Methods in Aquatic Bacteriology*. John Wiley & Sons. Chichester.

Brockman, E.R. (1967). Fruiting myxobacteria from the South Carolina coast. *Journal of Bacteriology*, 94, 1253-1254.

Buesing, N. & Gessner, M.O. (2002). Comparison of detachment procedures for direct counts of bacteria associated with sediment particles, plant litter and epiphytic biofilms. *Aquatic Microbial Ecology*, 27, 29-36.

Drews, G. (1974). *Mikrobiologisches Praktikum*, 2nd ed. Springer-Verlag. Berlin.

Dworkin, M. (ed.) (1999-2003) *The Prokaryotes: An Evolving Electronic Resource for the Microbiological Community*, 3rd edition, release 3.0, 21 May 1999, to release 3.15, 15 December 2003. Springer-Verlag. New York. http://link.springer-ny.com/link/service/books/10125/.

Head, I.M., Saunders, J.R. & Pickup, R.W. (1998). Microbial evolution, diversity, and ecology: A decade of ribosomal RNA analysis of uncultivated microorganisms. *Microbial Ecology*, 35, 1-21.

Hespell, R.B. & Bryant, M.P. (1981). The genera *Butyrivibrio, Succinivibrio, Lachnospira* and *Selenomonas*. In M.P. Starr, H. Stolp, H.G. Trüper, A. Balows & H.G. Schlegel (eds.), *The Prokaryotes: A Handbook on Habitats, Isolation, and Identification of Bacteria*, vol. 1 (pp. 1479–1494). Springer-Verlag. Berlin.

Overmann, J. (2003). Principles of enrichment, isolation, cultivation, and preservation of Prokaryotes. In M. Dworkin et al. (eds.), *The Prokaryotes: An Evolving Electronic Resource for the Microbiological Community*, 3rd edition, release 3.12, 28 March 2003. Springer-Verlag. New York. http://link.springer-ny.com/link/service/books/10125/.

Pelczar, M.J. & Chan, E.C.S. (1977). *Laboratory Exercises in Microbiology*, 4th ed. McGraw-Hill. New York.

Pfennig, N. & Trüper, H.G. (1989). Anoxygenic phototrophic bacteria. In: J.T. Staley, M.P. Bryant, N. Pfennig & J.G. Holt (eds.), *Bergey's Manual of Systematic Bacteriology*, vol. 3 (pp. 1635-1709). Williams and Wilkins. Baltimore.

Reichenbach, H. (1999). The order Cytophagales. In M. Dworkin et al. (eds.), *The Prokaryotes: An Evolving Electronic Resource for the Microbiological Community*, 3rd edition, release 3.0, 21 May 1999. Springer-Verlag. New York. http://link.springer-ny.com/link/service/books/10125/.

Reichenbach, H. & Dworkin, M. (1981). The order Cytophagales (with addenda on the genera *Herpetosiphon, Saprospira*, and *Flexithrix*). In: M.P. Starr, H. Stolp, H.G. Trüper, A. Balows & H.G. Schlegel (eds.), *The Prokaryotes: A Handbook on Habitats, Isolation, and Identification of Bacteria*, vol. 1 (pp. 356-379). Springer-Verlag. Berlin.

Rodina, A.G. (1972). *Methods in Aquatic Microbiology*. Translated, edited and revised by R.R. Colwell & M.S. Zambruski. University Park Press. Baltimore, and Butterworth. London.

Schneider, J. & Rheinheimer, G. (1988). Isolation methods. In: B. Austin (ed.), *Methods in Aquatic Bacteriology* (pp. 73-94). John Wiley & Sons. Chichester.

Stanier, R.Y. (1942). The *Cytophaga* group: a contribution to the biology of myxobacteria. *Bacteriological Reviews*, 6, 143–196.

Starr, M.P., Stolp, H., Trüper, H.G., Balows, A. & Schlegel, H.G. (eds.) (1981). *The Prokaryotes: A Handbook on Habitats, Isolation, and Identification of Bacteria*, vol. 1-2. Springer-Verlag. Berlin.

Trüper, H.G., Stackebrandt, E. & Schleifer, K.-H. (2001). Prokaryote characterization and identification. In: M. Dworkin et al. (eds.), *The Prokaryotes: An Evolving Electronic Resource for the Microbiological Community*, 3rd edition, release 3.6, June 22, 2001, Springer-Verlag. New York. http://link.springer-ny.com/link/service/books/10125/.

CHAPTER 30

EXTRACTION AND QUANTIFICATION OF ATP AS A MEASURE OF MICROBIAL BIOMASS

MANUELA ABELHO

Escola Superior Agrária, Instituto Politécnico de Coimbra, Bencanta, 3040-316 Coimbra, Portugal.

1. INTRODUCTION

ATP has been widely used to estimate total microbial biomass in a variety of systems (Karl 1980, Maltby 1992, Gessner 1997, Abelho 2001). The ubiquitous distribution of ATP in living cells, the rapid loss from dead cells, the fairly constant concentrations in microorganisms (Holm-Hansen & Karl 1978), and the ease of extraction and measurement have fostered the use of ATP as an indicator of living and active microbial biomass (McCarthy 1991).

ATP has been shown to exhibit a significant positive correlation with ergosterol, a specific indicator of fungal biomass (Chapter 25), during leaf decomposition in streams (Suberkropp et al. 1993). This observation suggests that ATP may be interpreted as an indicator of fungal biomass in situations where bacteria and other microorganisms do not contribute significantly to the total ATP pool (Suberkropp et al. 1993). That situation may well be encountered on leaves decomposing in streams, which are often pervasively colonized by fungi and much less so by bacteria (Baldy et al. 1995). Thus, some researchers have used ATP to assess fungal biomass associated with decomposing leaves in streams, rather than total microbial biomass (Rosset et al. 1982, Suberkropp et al. 1983, Suberkropp 1991).

Numerous methods are available to extract ATP from microbial cells. Important requisites for efficient extraction are rapid cell death and lysis, complete ATP release, complete and irreversible inactivation of enzymes, long-term stability of extracted ATP (Karl 1980), and lack of inhibition of the firefly reaction (Gregg 1991). Although many different extraction procedures have been proposed, they generally fall into one of two categories: extraction with boiling buffers or with cold acids. Extraction with boiling buffers has proven efficient for fungal spores (Rakotonirainy et al. 2003) and freshwater zooplankton (Amyot et al. 1992);

M.A.S. Graça, F. Bärlocher & M.O. Gessner (eds.), Methods to Study Litter Decomposition: A Practical Guide, 223 – 230.

although boiling buffer extracted more ATP, cold H_2SO_4 gave more consistent results (Amyot et al. 1992). Extraction efficiencies of ATP in microorganisms associated with non-living organic matter (e.g. decomposing leaves in streams) may be low with cold acids leading to higher recovery rates than boiling buffers (e.g. Holm-Hansen & Karl 1978, Karl 1980).

Methods for the quantification of ATP include high-performance liquid chromatography (HPLC) with phosphate buffer as the mobile phase and UV detection, and ion exchange chromatography, also with UV detection (Ally & Park 1992, Maguire et al. 1992). However, the firefly luciferin-luciferase bioluminescence method is the most rapid, sensitive, and reproducible assay. The method is based on the bioluminescent reaction catalysed by the firefly luciferase:

$$\text{MgATP} + \text{luciferin} \xrightarrow{\text{luciferase}} \text{PP}_i + \text{AMP - luciferin}$$
$$\xrightarrow{O_2} \text{oxyluciferin} + \text{AMP} + CO_2 + \text{light}$$

(30.1)

Maximum intensity of the emitted light is at 562 nm. The optimum pH for the firefly luciferin-luciferase reaction is 7.75 and the optimum temperature is about 25 °C (DeLuca 1976). Under these conditions, and at low ATP concentrations (i.e. within the "linear" portion of the Michaelis-Menten curve), light intensity is directly proportional to ATP concentration. The great sensitivity of the method allows detection of 0.1 fmol ATP (50 fg) or even less (Gregg 1991).

The emitted light is usually measured by integration of the light signal during a specified period or by light-intensity measurements on specific phases of the curve, usually at peak intensity (Lundin & Thore 1975). The integration of the light flux for a set time period has three major advantages: (1) increased sensitivity, (2) ease and reliability of mixing, and (3) no need to depend upon a peak-height response (Holm-Hansen & Karl 1978). An integration period of 30 s has shown a linear relationship between ATP concentrations and light emission over at least three orders of magnitude and has been used to quantify ATP associated with decomposing leaves in streams (Suberkropp et al. 1983).

The most practical and convenient way to relate the measured light flux to ATP concentrations is to use a series of internal standards (Holm-Hansen & Karl 1978), i.e. the addition of known amounts of ATP to the sample at various steps during the analytical procedure. Internal standardization also compensates for potential interferences with bioluminescence by several components other than ATP, which typically occur in complex environmental samples (Lundin & Thore 1975, Holm-Hansen & Karl 1978).

It is often convenient to express ATP levels in terms of total microbial biomass carbon or dry mass. The factor most commonly used to convert measured ATP values to total biomass carbon is 250 (Holm-Hansen & Karl 1978, Karl 1980). Although this conversion factor is the mean established from data on seven strains of marine bacteria and 30 species of unicellular marine algae, it has been systematically applied to diverse systems without further verification of its general validity (Karl 1980). However, an average ATP concentration of 1.8 mg g^{-1} dry mass was found in several fungal species colonizing decomposing leaves in streams (Suberkropp 1991, Suberkropp et al. 1993). In accordance with the average calculated by Karl (1980), this concentration results in a conversion factor of exactly 250 if fungal biomass carbon is 45% of dry mass, which is a very reasonable assumption (e.g. Baldy et al. 1995).

The aim of the procedure described here is to quantify ATP in decomposing leaves, as an estimate of microbial colonization. ATP is quantified by bioluminescence after extraction in cold sulphuric acid and the buffer HEPES (e.g. Holm-Hansen & Karl 1978, Karl 1980, Suberkropp et al. 1983). Some ATP values from the literature are shown in Table 30.1.

Table 30.1. ATP concentrations of decomposing leaf litter in streams. AFDM = ash-free dry mass.

Maximum ATP concentration (nmol g^{-1} AFDM) *	Leaf species	Stream type	Reference
26—257	Liriodendron tulipifera	Softwater	1, 2, 3
99—330	Liriodendron tulipifera	Hardwater	1, 2, 3
150	Picea abies	Softwater	4
250	Quercus petraea	Softwater	4
300	Larix decidua	Softwater	4
300	Larix decidua	Hardwater	4
300	Picea abies	Hardwater	4
500	Quercus petraea	Hardwater	4
500	Castanea sativa	Softwater	5
750	Hura crepitans	Tropical hardwater	5

* Multiply value by 0.6052 to convert nmol to µg. 1 = Suberkropp (1991); 2 = Suberkropp et al. (1993); 3 = Suberkropp & Chauvet (1995); 4 = Rosset et al. (1982); 5 = Abelho (1999).

2. EQUIPMENT, CHEMICALS AND SOLUTIONS

2.1. Equipment and Material

2.1.1. ATP Extraction
- Forceps
- Pipettes (0.1 ml and 5 ml)
- Homogenizer (Polytron)
- Centrifuge tubes (50 ml)
- Refrigerated centrifuge (10,000 g)
- Sterile filters (0.2 µm) and filter holders
- Syringe (20 ml)

- Stirring plate and bars
- pH meter
- Glassware (20 ml beakers, 20 ml volumetric cylinders)

2.1.2. ATP Quantification

- Vortex
- Reaction vessels: polypropylene test tubes (1.6 ml; 8 × 50 mm) with hydrophobic inner surfaces (e.g. Turner Designs)
- Pipettes (200, 100, 20, and 10 µl)
- Luminometer (e.g. Turner Designs TD-20/20), consisting of a reaction chamber mounted vertically, such that the light emitted from the reaction vessel inside the chamber reaches the photosensitive surface of a photomultiplier tube.

2.2. Chemicals

2.2.1. ATP extraction

- ATP (Adenosine-5'-triphosphate: $C_{10}H_{14}N_5O_{13}P_3Na_2 \cdot 3H_2O$; e.g. Boehringer Mannheim)
- HEPES buffer (N-[2-Hydroxyethyl]piperazine-N'-[2-ethanesulfonic acid]: $C_8H_{17}N_2O_4SNa$; e.g. Sigma)
- Oxalic acid (Ethanedioic acid: $C_2H_2O_4 \cdot 2H_2O$)
- Sulphuric acid, 95-97% (H_2SO_4)
- Ammonium hydroxide (NH_4OH)
- Hydrochloric acid, 37% (HCl)
- Ultrapure water (e.g. Seral Pur PRO 90 CN)

2.2.2. ATP Quantification

- Firelight[®] (Analytical Luminescence Laboratory, Ann Arbor, MI, USA): a highly purified mixture of luciferase, bovine serum albumine and luciferin
- ATP (Adenosine-5'-triphosphate: $C_{10}H_{14}N_5O_{13}P_3Na_2 \cdot 3H_2O$; e.g. Boehringer Mannheim)
- HEPES buffer (N-[2-Hydroxyethyl]piperazine-N'-[2-ethanesulfonic acid]: $C_8H_{17}N_2O_4SNa$; e.g. Sigma)

2.3. Solutions

- Solution 1: 100 µM ATP prepared in ultrapure water. Store at –20 °C in Eppendorf tubes for up to 2 months
- Solution 2: 0.05 M HEPES prepared in ultrapure water and adjusted to pH 7.5 with HCl; store at 4 °C for up to 2 months

- Solution 3: 0.6 M sulphuric acid containing 8 g l^{-1} oxalic acid; store at 4 °C for up to 2 months
- Solution 4: 1 μM ATP prepared in ultrapure water; store at –20 °C in Eppendorf tubes for up to 2 months
- Solution 5: Luciferin-luciferase solution (dissolve the contents of one sealed vial in 5 ml of HEPES buffer). Keep at 4 °C, protected from light, for up to 15 days. Freshly prepared enzyme emits some light without the addition of ATP; however, the endogenous light usually falls below detectable levels upon storage overnight in the refrigerator (Lundin & Thore 1975). Unless the enzyme solution is used on several successive days; store at –20 °C for up to 2 months

3. EXPERIMENTAL PROCEDURES

3.1. Sample Preparation

1. Collect leaves decomposing in stream and transport to laboratory in ice-chest.
2. Clean the leaves from adhering debris and invertebrates.
3. Cut duplicate sets of leaf discs with a cork borer (suggested diameter 12—14 mm).
4. Oven-dry (40—50 °C, 72 h) one set of leaf discs and weigh to the nearest 0.1 mg to determine dry mass. Ash (500 °C, 5 h), weigh to the nearest 0.1 mg and subtract from dry mass to determine ash-free dry mass (AFDM).
5. Use the second set to determine ATP concentration of the sample.

3.2. ATP Extraction

1. Homogenize leaf discs with Polytron homogenizer (position 30) for 15 s in 5 ml of cold Solution 3 and 5 ml of Solution 2.
2. Centrifuge for 20 min at 4 °C and 10,000 g.
3. Filter with 0.2 μm pore size filters in filter holders.
4. Adjust pH of the supernatant to 7.0—7.5 with ammonium hydroxide, using a magnetic stirrer and pH meter.
5. Note the final volume and store extract in 20-ml scintillation vials.
6. Freeze at –20 °C until ATP is measured.
7. Carry a parallel set of sample discs, known to contain no living organisms but spiked with a known amount of ATP (suggested amount 50 μl of Solution 1, i.e., 5 nmol) through the whole procedure to determine recovery of the extracted ATP.

3.3. ATP Quantification

1. Set up the luminometer according to the specific instruction of the instrument.
2. Choose an integration period of 30 s and, if possible, choose a double measurement mode.

3. Add 50 μl Firelight®, 130 μl HEPES buffer and 20 μl of the sample in a reaction vessel.
4. Vortex lightly or mix by hand.
5. Insert the reaction vessel into the chamber and record light intensity.
6. For internal calibration, remove the reaction vessel as fast as possible, and add 2 μl of the standard 1 μM ATP solution (Solution 4).
7. Record light intensity.
8. Repeat all steps once and calculate average ATP concentration of the extract.

3.4. Calculations

1. The amount of ATP present in the assay sample is obtained with the following formula:

$$ATP = \frac{A}{(B-A)} \cdot \text{amount of added ATP} \tag{30.2}$$

where A is the light emission recorded before the addition of the internal ATP standard, and B is the light emission after the addition of the internal standard.
2. To determine total amount of ATP in the sample, multiply the amount of ATP measured in 20 μl of the sample extract and divide by the volume analyzed:

$$ATP \text{ in sample} = \frac{ATP \text{ in 20 μl} \cdot \text{sample volume (ml)}}{20 \text{ μl} \cdot 1000} \tag{30.3}$$

3. To determine extraction efficiency, use the relation between the known amount of ATP added to the control sample before extraction and the measured ATP:

$$\text{Extraction efficiency} = \frac{\text{Measured ATP in control sample}}{\text{Amount of ATP added before extraction}} \tag{30.4}$$

4. To account for losses of ATP during sample analysis, divide the measured ATP values by the extraction efficiency (Equation 19.4).
5. To convert nmol of ATP to μg multiply by 0.6052.

4. REFERENCES

Abelho, M. (1999). *Once upon a time a leaf: from litterfall to breakdown in streams.* PhD Thesis. University of Coimbra, Coimbra, Portugal.

Abelho, M. (2001). From litterfall to breakdown in streams: a review. *TheScientificWorld,* 1, 656-680.

Ally, A. & Park, G. (1992). Rapid-determination of creatine, phosphocreatine, purine-bases and nucleotides (ATP, ADP, AMP, GTP, GDP) in heart biopsies by gradient ion-pair reversed-phase liquid-chromatography. *Journal of Chromatography – Biomedical Applications,* 575, 19-27.

Amyot, M., Pinelalloul, B., Bastien, C., Methot G., Blaise, C., Vancoillie, R. & Thellen, C. (1992). Firefly assay of ATP from freshwater zooplankton – comparison of extraction methods. *Environmental Toxicology and Water Quality,* 7, 295-311.

Baldy, V., Gessner, M.O. & Chauvet, E. (1995). Bacteria, fungi and the breakdown of leaf litter in a large river. *Oikos,* 74, 93-102.

DeLuca, M. (1976). Firefly luciferase. *Advances in Enzymology,* 44, 37-68.

Gessner, M.O. (1997). Fungal biomass, production and sporulation associated with particulate organic matter in streams. *Limnética,* 13, 33-44.

Gregg, C.T. (1991). Bioluminescence in clinical microbiology. In W.H. Nelson (ed.), *Physical Methods for Microorganisms Detection* (pp. 1-28). CRC Press. Boca Raton, Florida.

Holm-Hansen, O. & Karl, D.M. (1978). Biomass and adenylate energy charge determination in microbial cell extracts and environmental samples. *Methods in Enzymology,* 57, 73-85.

Karl, D.M. (1980). Cellular nucleotide measurements and applications in microbial ecology. *Microbiological Reviews,* 44, 739-796.

Lundin, A. & Thore, A. (1975). Analytical information obtainable by evaluation of the time course of firefly bioluminescence in the assay of ATP. *Analytical Biochemistry,* 66, 47-63.

Maguire, M.H., Szabo, I. Slegel, P. & King, C.R. (1992). Determination of concentrations of adenosine and other purines in human term placenta by reversed-phase high-performance liquid-chromatography with photodiode-array detection – evidence for pathways of purine metabolism in the placenta. *Journal of Chromatography – Biomedical Application,* 575, 243-253.

Maltby, L. (1992). Heterotrophic microbes. In P. Calow & G.E. Petts (eds.), *The Rivers Handbook: Hydrological and Ecological Principles* (pp.165-194). Blackwell Science. Oxford.

McCarthy, B. J. (1991). The use of ATP measurements in biodeterioration studies. In W.H. Nelson (ed.) *Physical Methods for Microorganisms Detection* (pp. 129-138). CRC Press.Boca Raton, Florida.

Rakotonirainy, M.S., Heraud, C. & Lavedrine, B. (2003). Detection of viable fungal spores contaminant on documents and rapid control of the effectiveness of an ethylene oxide disinfection using ATP assay. *Luminescence,* 18, 113-121.

Rosset, J., Bärlocher, F. & Oertli, J.J. (1982). Decomposition of conifer needles and deciduous leaves in two black forest and two Swiss Jura streams. *Internationale Revue der gesamten Hydrobiologie,* 67, 695-711.

Suberkropp, K. (1991). Relationships between growth and sporulation of aquatic hyphomycetes on decomposing leaf litter. *Mycological Research,* 95, 843-850.

Suberkropp, K. & Chauvet, E. (1995). Regulation of leaf breakdown by fungi in streams: influences of water chemistry. *Ecology,* 76, 1433-1445.

Suberkropp, K., Arsuffi, T.L. & Anderson, J.P. (1983). Comparison of degradative ability, enzymatic activity, and palatability of aquatic hyphomycetes grown on leaf litter. *Applied and Environmental Microbiology,* 46, 237-244.

Suberkropp, K., Gessner, M.O. & Chauvet, E. (1993). Comparison of ATP and ergosterol as indicators of fungal biomass associated with decomposing leaves in streams. *Applied and Environmental Microbiology,* 59, 3367-3372.

CHAPTER 31

RESPIROMETRY

MANUEL A.S. GRAÇA[1] & MANUELA ABELHO[2]

[1]*Departamento de Zoologia, Universidade de Coimbra, 3004-517 Coimbra, Portugal.* [2]*Escola Superior Agrária, Instituto Politécnico de Coimbra, Bencanta, 3040-316 Coimbra, Portugal.*

1. INTRODUCTION

Community respiration is a measure of biological activity, reflecting the microbial use of organic matter and, therefore, the functional significance of microbes in decomposition. Respiration is also an indicator of energy allocated to metabolic processes of invertebrates. In terrestrial systems, respiration rates are usually calculated from CO_2 fluxes measured with infrared gas analyzers (e.g. Eriksen & Jensen 2001, Kuehn & Suberkropp 2000). In aquatic ecosystems, in contrast, respiration rates both of microorganisms associated with decomposing leaves and of invertebrates are generally determined by measuring oxygen consumption (e.g. Ramirez et al. 2003, Rier et al. 2002, Table 31.1).

Respiration can be measured in closed or open systems. The simplest closed system consists of a water-filled flask with no head space and well sealed to prevent gas exchange with the environment (e.g. Iversen 1979). Water circulation can be achieved by magnetic stirrers, isolated from the leaf material by a wire mesh (Quinn et al. 2000) or by a pump creating a unidirectional flow in a recirculating chamber (Royer & Minshall 2001). Oxygen concentrations can be measured continuously with an oxygen electrode (e.g. Kominková et al. 2000) or at intervals using either probes or the Winkler method (e.g. Hill et al. 2000). Another system frequently used in both soil and aquatic environments is the Gilson differential respirometer, which measures respiration as the change in pressure in respiration chambers due to the production or consumption of gases (e.g. Tank & Webster 1998, Simon & Benfield 2001).

As an alternative to closed systems, respiration may be measured in a flow-through system, with a leaf sample placed inside a chamber through which oxygenated water flows. Respiration is calculated as the difference in oxygen concentration in the incoming and outflowing water multiplied by the flow rate. The flow-through method described here was adapted from Wrona & Davies (1984) and

M.A.S. Graça, F. Bärlocher & M.O. Gessner (eds.), Methods to Study Litter Decomposition: A Practical Guide, 231 – 236.

has been applied in a variety of studies aiming at assessing respiration rates of leaf-shredding invertebrates (e.g. Naylor et al. 1989, Graça et al. 1993).

Table 31.1. Respiration rates of microorganisms associated with decomposing plant litter in streams. AFDM = ash-free dry mass; n.d. = not determined.

Respiration ($\mu g\ O_2\ h^{-1}$ mg^{-1} AFDM)	Sample[b]	Method	Temp. (°C)	Ref.
0.3	*Salix* sp.	Closed chambers	10	1
≤0.5	*Acer rubrum* and *Rhododendron maximum*	Closed chambers in stream	4—15	2
0.75—1.5	*Cornus stolonifera, Populus tremuloides* and *Betula occidentalis*	*In situ* recirculating chambers	Ambient	3
1.90—2.49	*Ficus insipida*	Closed recirculating chambers	24—27	4
200—440[a]	*Populus tremuloides*	BOD flaks with current	15	5
0.05—0.65	*Sericostoma personatum*	Closed flasks	n.d.	6
0.51—4.99	*Sericostoma vittatum*	Flow-through system	15	7
0.93—2.07	*Lepidostoma unicolor*	Gilson respirometer	5—15	8
0.44—2.22[a]	*Asellus aquaticus*	Gilson respirometer	n.d.	9

[a] Converted from dry mass (DM) values assuming that AFDM is 90% of DM. [b] Plant names refer to decomposing leaves. 1 = Niyogi et al. (2002); 2 = Gulis & Suberkropp (2003); 3 = Royer & Minshall (2001); 4 = Ramirez et al. (2003); 5 = Rier et al. (2002); 6 = Grafius & Anderson (1980); 7 = Feio & Graça (2000); 8 = Iversen (1979); 9 = Adock (1982);

2. EQUIPMENT, CHEMICALS AND SOLUTIONS

2.1. Equipment and Materials

- Stream incubated leaves or invertebrates
- Constant temperature room (e.g. 15 °C)
- Oxygen meter, oxygen electrode and measuring cell (approx. 0.1 ml)
- Peristaltic pump connected to multiple silicon tubes (e.g. 0.38 mm bore)
- Respiration chambers: 5 ml glass syringe with the plunger replaced by a rubber/silicon stopper. The stopper should be pierced by a blunted hypodermic syringe with a Luer lock.
- Micro-syringe (0.5 ml)
- Water tank (capacity 5 l)
- Measuring cylinder (5 ml)
- Aeration system (e.g. aquarium pump or pressurized air)
- Cork borer (e.g. 1 cm diameter)

- Drying oven (set at 40—50 °C)
- Analytical balance

2.2. Chemicals and Solutions

- Filtered stream water
- Oxygen-free calibration solution for the oxygen meter: Sodium borate (3.8 g l^{-1}) saturated with crystalline sodium sulphite.

3. EXPERIMENTAL PROCEDURES

1. Calibrate the oxygen meter with oxygen-saturated water and the oxygen-free solution.
2. With a cork borer punch out discs from leaf samples collected from a stream. Insert 5—8 leaf discs in each respiration chamber.
3. Fill the respiration chamber with aerated water. Pump filtered aerated stream water from the tank into the respiration chambers via tubes connected to the Luer lock of the hypodermic needle (Fig 31.1).
4. After complete renewal of the syringe volume, take water sample from the respiration chambers with a 0.5-ml micro-syringe and transfer to a measuring cell where an oxygen electrode is inserted. Read the oxygen concentration after a 30-s stabilization period.
5. Initial oxygen concentration in the water entering the chambers can be measured between readings by taking samples from a control respiration chamber or from a valve at the inflow to chambers.
6. The oxygen meter should be checked and, if necessary, re-calibrated between readings to compensate for drift.
7. Oxygen consumption is given by the difference in oxygen concentrations in the outflow of a chamber containing a biological sample, $[O_2]_b$, and the outflow of an empty control chamber, $[O_2]_c$. The rate of water flow through the respiration chambers (v) can be measured by collecting the outflowing water in a 5-ml measuring cylinder for 10—20 min.
8. Since high flow rates decrease the sensitivity of the measurement, and low levels of oxygen can affect respiration, the peristaltic pump is best adjusted to deliver a rate at which oxygen concentration of the water leaving the chamber is approximately 70% of the concentration of the inflowing water at saturation level.
9. After at least three separate readings per chamber, transfer the leaf discs to aluminium pans and dry in the oven to constant weight (typically for 2 days). Determine the dry mass (DM) of leaf discs to the nearest 0.01 mg.
10. Respiration rates (R) can be expressed as mg O_2 consumed per mg leaf dry mass per hour:

$$R = \frac{[O_2]_b - [O_2]_c \cdot v}{DM \cdot t} \qquad (31.1)$$

Alternatively, respiration rates may be expressed per leaf area (A) as mg O_2 consumed per cm^2 per hour:

$$R = \frac{[O_2]_b - [O_2]_c \cdot v}{A \cdot t} \qquad (31.2)$$

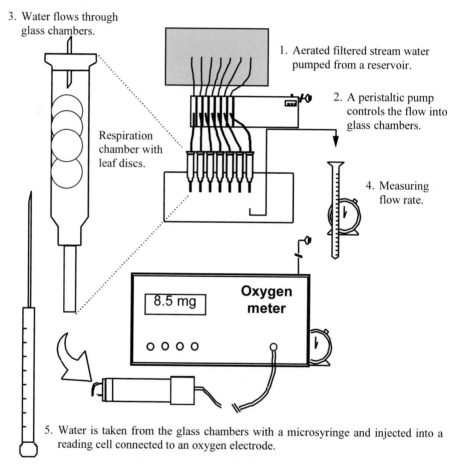

3. Water flows through glass chambers.

1. Aerated filtered stream water pumped from a reservoir.

2. A peristaltic pump controls the flow into glass chambers.

Respiration chamber with leaf discs.

4. Measuring flow rate.

Oxygen meter

8.5 mg

5. Water is taken from the glass chambers with a microsyringe and injected into a reading cell connected to an oxygen electrode.

Figure 31.1. Respiration system (modified from Graça, 1990)

4. FINAL REMARKS

Before measurements are taken, leaf discs must be acclimated to the chamber conditions to ensure accurate readings. Acclimation may take 1.5 h or even longer. Moreover, readings should be made only after replacement of all water in the respiration chambers.

Since oxygen solubility in water and respiration are temperature dependent, and activities of consumers are also affected by photoperiod (e.g. Adock 1982, Fuss & Smock 1996, Feio & Graça 2000), measurements must be made at a constant temperature and the photoperiod must be controlled.

Multiply concentrations given in ml l^{-1} by 1.33 to obtain values in mg l^{-1}. Multiply concentrations given in ml l^{-1} by 4.16 10^{-4} to obtain mol O_2 (standard conditions of 20 °C and 1 MPa).

4. REFERENCES

Adock, J.A. (1982). Energetics of a population of *Asellus aquaticus* (Crustacea, Isopoda): respiration and energy budgets. *Freshwater Biology*, 12, 257-269.

Eriksen, J. & Jensen, L.S. (2001). Soil respiration, nitrogen mineralization and uptake in barley following cultivation of grazed grassland. *Biology and Fertility of Soils*, 33, 139-145.

Feio, M.J. & Graça, M.A.S. (2000). Food consumption by the larvae of *Sericostoma vittatum* (Trichoptera), an endemic species of the Iberian Peninsula. *Hydrobiologia*, 439, 7-11.

Fuss, C.L. & Smock, L.A. (1996). Spatial and temporal variation of microbial respiration rates in a blackwater stream. *Freshwater Biology*, 36, 339-349.

Grafius, E. & Anderson, N.H. (1980). Population dynamics and the role of two species of *Lepidostoma* (Trichoptera: Lepidostomatidae) in an Oregon coniferous forest stream. *Ecology*, 61, 808-816.

Graça, M.A.S. (1990). Observations on the feeding biology of two stream-dwelling detritivores: *Gammarus pulex* (L.) and *Asellus aquaticus* (L.). PhD. Thesis, University of Sheffield, U.K.

Graça, M.A.S., Maltby, L. & Calow, P. (1993). Importance of fungi in the diet of *Gammarus pulex* and *Asellus aquaticus*: II. Effects on growth, reproduction and physiology. *Oecologia*, 96, 304-309.

Gulis, V. & Suberkropp, K. (2003). Leaf litter decomposition and microbial activity in nutrient-enriched and unaltered reaches of a headwater stream. *Freshwater Biology*, 48, 123-134.

Hill, B.H., Hall, R.K., Husby, P., Herlihy, A.T. & Dunne, A.M. (2000). Interregional comparisons of sediment microbial respiration in streams. *Freshwater Biology*, 44, 213-222.

Iversen, T.M. (1979). Laboratory energetics of larvae of *Sericostoma personatum* (Trichoptera). *Holarctic Ecology*, 2, 1-5.

Komínková, D., Kuehn, K.A., Buesing, N., Steiner, D. & Gessner, M.O. (2000). Microbial biomass, growth, and respiration associated with submerged litter of *Phragmites australis* decomposing in a littoral reed stand of a large lake. *Aquatic Microbial Ecology*, 22, 271-282.

Kuehn, K.A. & Suberkropp, K.. 1998. Diel fluctuations in rates of CO_2 evolution from standing dead leaf litter of the emergent macrophyte *Juncus effusus*. *Aquatic Microbial Ecology*, 14, 171-182.

Naylor, C., Maltby, L. & Calow, P. (1989). Scope for growth in *Gammarus pulex*, a freshwater benthic detritivore. *Hydrobiologia*, 188/189, 517-523.

Niyogi, D.K., McKnight, D.M. & Lewis Jr., W.M. (2002). Fungal communities and biomass in mountain streams affected by mine drainage. *Archiv für Hydrobiologie*, 155, 255-271

Quinn, J.M., Burrell, G.P. & Parkyn, S.M. (2000). Influences of leaf toughness and nitrogen content on in-stream processing and nutrient uptake by litter in a Waikato, New Zealand, pasture stream and streamside channels. *New Zealand Journal of Marine Freshwater Resaerch*, 34, 353-271.

Ramirez, A., Pringle, C.M., & Molina, L. (2003) Effects of stream phosphorus levels on microbial respiration. *Freshwater Biology*. 48, 88-97.

Rier, S.T., Tuchman, N.C., Wetzel R.G. & Teeri, J.A. (2002). Elevated-CO_2-induced changes in the chemistry of quaking aspen (*Populus tremuloides* Michaux) leaf litter: subsequent mass loss and microbial response in a stream ecosystem. *Journal of the North American Benthological Society*, 21, 16-27.

Royer, T.V. & Minshall, G.W. (2001). Effects of nutrient enrichment and leaf quality on breakdown of leaves in a hardwater stream. *Freshwater Biology*, 46, 603-610.

Simon, K.S. & Benfield, E.F. (2001). Leaf and wood breakdown in cave streams. *Journal of the North American Benthological Society*, 20, 550-563.

Tank, J.L. & Webster, J.R. (1998). Interaction of substrate and nutrient availability on wood biofilm processes in streams. *Ecology*, 79, 2168-2179.

Wrona, F. & Davies, R. (1984). An improved flow-through respirometer for aquatic macroinvertebrate bioenergetic research. *Canadian Journal of Fisheries and Aquatic Sciences*, 41, 380-385.

PART 4

ENZYMATIC CAPABILITIES

CHAPTER 32

EXTRACTELLULAR FUNGAL
HYDROLYTIC ENZYME ACTIVITY

SHAWN D. MANSFIELD

Canada Research Chair in Wood and Fibre Quality, Department of Wood Science, University of British Columbia, Vancouver, B.C., Canada, V6T 1Z4.

1. INTRODUCTION

Plant polysaccharides are the most abundant organic polymers in the biosphere. In their natural environment, they are used, degraded and re-mineralized primarily via the biological activities of bacteria and fungi. Microorganisms produce a battery of extracellular hydrolytic and oxidative enzymes that depolymerize cellulose, hemicelluloses, lignin and other polymers into smaller, more readily utilizable compounds. These breakdown products pass through both the cell wall and plasma membrane, and serve as energy sources and/or precursors in cell biosynthesis.

1.1. Cellulose Degradation

Cellulose is the most abundant biopolymer on earth, found primarily as a structural component of the cell wall of plants and marine algae. The gross physical structure and morphology of cellulose consists of long, unbranched homopolymers of D-glucose units linked by β-1,4-glycosidic bonds to form a linear chain of over 10,000 glucose residues (Hon & Shiraishi 1991). Individual glucan chains adhere to each other by hydrogen bonding and van der Waals forces and form insoluble networks. The secondary and tertiary structures of native cellulose are complex, and may vary significantly depending on the source and biosynthetic machinery that produced the polymer (i.e. plant or bacteria). Furthermore, the cellulose polymers of higher plants are intricately associated with lignin and hemicellulose moieties resulting in even more complex morphologies.

Primary cellulose degradation results from either chemical or enzymatic hydrolysis of the polymer into oligomeric and monomeric soluble sugars. Due to the inherent insolubility and physical complexity of cellulose moieties, several different enzymes are needed for complete solubilization (Mansfield et al. 1999).

239

M.A.S. Graça, F. Bärlocher & M.O. Gessner (eds.), Methods to Study Litter Decomposition: A Practical Guide, 239–248.
© 2005 *Springer.* Printed in The Netherlands.

The current understanding is that enzyme-mediated hydrolysis of native cellulose results primarily from the synergistic interaction of extracellular β-1,4-endoglucanases and β-1,4-exoglucanases (cellobiohydrolases) to yield cello-oligosaccharides such as cellobiose, which are subsequently cleaved to glucose by β-glucosidase. The activities of endo- and exo-glucanases are synergistic (Mansfield et al. 1999). The general mechanism suggests that the endoglucanases produce free chain ends on the cellulose surface for the cellobiohydrolases to act upon. However, synergy has also been observed between different types of cellobiohydrolases (Nidetzky et al. 1993), as well as between two endoglucanases (Gübitz et al. 1998, Mansfield et al. 1998). Although all cellulolytic enzymes have similar bond specificities (β-1,4), important functional differences are found in their mode of action towards solid substrates.

Generally, the activity of endoglucanases is assayed with a water-soluble substrate, such as carboxymethylcellulose (CMC) or phosphoric acid-swollen cellulose. The assay quantifies the amount of reducing sugars released from the substrate by the interaction with the enzyme (Ghose 1987). In contrast, exoglucanases differ substantially in their substrate specificity, and are capable of solubilizing crystalline cellulose substrates, such as Avicel, filter paper or cotton. Their activity is also usually measured by the amount of reducing ends generated (Ghose 1987). An alternative method for measuring activity uses a chromophoric disaccharide derivative and a homologous series of 4-methylumbelliferyl glycosides of cello-oligosaccharides (van Tilbeurgh et al. 1982, Chapter 35).

β-Glucosidases catalyze the hydrolysis of terminal, non-reducing β-D-glucose residues from β-D-glucosides, including cellobiose and cello-oligosaccharides. In some cases mixed oligosaccharides consisting of mannose and glucose serve as substrates. In the enzymatic conversion of cellulose, it is important that the level of β-glucosidase is in excess, as cellobiose has an inhibitory effect on the cellobiohydrolases (Mansfield et al. 1999).

1.2. Hemicellulose Degradation

Hemicelluloses are low-molecular weight heteropolymic polysaccharides constructed from a number of different residues, the most common of which are D-xylose, D-mannose, D-galactose, D-glucose, L-arabinose, D-rhamnose, D-galacturonic acid, D-glucuronic acid and 4-O-methyl-D-glucuonic acid (Fengel & Wegener 1983, Sjöström 1993). The complexity and chemical nature of the hemicelluloses varies both between cell types and species.

1.2.1. Xylan
The main xylan-derived hemicelluloses are polysaccharides with a backbone of 1,4-linked β-D-xylopyranosyl units, substituted at the carbon 2 and 3 positions. The extent of substitution is dependant on origin (Sjöström 1993): deciduous and coniferous-derived xylans carry 4-O-methylglucuronic acid and L-arabinofuranosyl

side groups, respectively, while xylans from annual plants may contain only the latter or both side groups. Furthermore, xylans derived from deciduous trees are acetylated, whereas coniferous-derived xylans are not (Sjöström 1993). Xylans from annual plants may, in addition to acetyl groups, carry esterified phenolic hydroxycinnamic acids such as feruloyl and *p*-coumaroyl moieties (Grabber et al. 2000).

The crucial enzyme for xylan depolymerization is endo-β-1,4-xylanase, which preferentially attacks the main xylan chain, generating non-substituted and branched or esterified oligosaccharides. The branching substituents are liberated by corresponding glycosidases or esterases (debranching or accessory enzymes): α-L-arabinofuranosidases and α-glucuronidase. Finally, acetic acid, ferulic acid and *p*-coumaric acid residues can be liberated from the xylan by corresponding xylan esterases. β-xylosidase liberates D-xylose from the non-reducing end of xylo-oligosaccharides.

1.2.2. Mannan

Mannan-based hemicelluloses are substituted heteropolysaccharides that are widespread in both deciduous and coniferous trees. Their designation is largely dependent on the constituent monomers comprising the backbone and the side chains, and can be divided into (1) pure mannans, (2) glucomannans, (3) galactomannans and (4) galactoglucomannans.

The biodegradation of β-mannans occurs by the synergistic action of endo-1,4-β-mannanases, β-D-mannosidases, β-D-glucosidases, α-D-galactosidases and acetyl mannan esterases (Tenkanen et al. 1993). Endo-β-mannanases cleave polymeric mannans as well as mannooligosaccharides, usually with a degree of polymerization greater than three. Some endomannanases also cleave β-1,4 linkages between mannose and glucose in glucomannans. The degree of substitution and the distribution of the side groups significantly influence the overall capacity for endomannanase to catalyze the degradation of β-1,4 linkages. Thus, the combined actions of endomannanases and accessory enzymes such as α-galactosidase and acetyl esterase are required for total degradation of galactoglucomannan (Tenkanen et al. 1993).

β-Mannosidase catalyze the hydrolysis of terminal, non-reducing β-D-mannose residues in mannans, heteromannans and mannooligosaccharides. Some β-mannosidases also cleave the 1,4-β-mannose-glucose linkages in glucomannans. β-mannosidases occur in a wide range of plant and animal tissues and in many microorganisms (Gübitz et al. 1996).

1.3. Measuring Enzyme Activities

The use of purified enzymes is essential to determine substrate specificities of individual enzymes and to elucidate molecular mechanisms of catalysis. However, simplified assays exist for the determination of each of the general classes of extracellular hydrolytic (cellulolytic and hemicellulolytic) enzymes secreted by microorganisms.

This chapter presents three types of methods to quantify the major hydrolytic enzymes in fungal cultures, but does not include procedures to determine the debranching enzymes required for total cell wall carbohydrate degradation: (1) The determination of β-1,4-endoglucanases, β-1,4-endoxylanase and β-1,4-endomannanase (on any mannan-based substrate) follows a variation of Bailey et al. (1992), with appropriate substitution for substrates and corresponding standards. For example, β-1,4-endoglucanases activity is analyzed on carboxymethylcellulose using glucose as a standard. (2) Filter paper activity is a good measure of total cellulase activity. Since exoglucanases are required for the solubilization of crystalline cellulose, this method is also a relatively good indicator of the presence of cellobiohydrolase; however, it does not specifically quantify exoglucanases (Ghose, 1987). (3) Finally, β-glucosidase, β-xylosidase, β-mannosidase activities are quantified based on an assay by Ghose (1987), with appropriate substitution for substrates and corresponding standards. For example, β-glucosidase activity is determined using p-nitrophenyl-β-glucoside and glucose.

2. EQUIPMENT, CHEMICALS AND SOLUTIONS

2.1. Equipment and Material

- Analytical balance
- Cooled centrifuge (4 °C, 20000 g)
- Shaking incubator (20 °C)
- Boiling water bath
- Water bath (50 °C)
- Water bath (20 °C)
- pH meter
- Magnetic stirrer
- Spectrophotometer
- Vortex
- Laboratory timer or stopwatch
- Adjustable micropipettors (0.2—1.8 ml)
- Petri plates
- Erlenmeyer flasks
- Centrifuge tubes
- Cuvettes (disposable are suitable)
- Test tubes (15 ml)
- Test tube rack
- Filter paper (Whatman No. 1)

- Fungal isolates maintained on 1% malt agar plates at 15—20 °C (to isolate aquatic hyphomycetes, see Chapter 21)
- Sterilized leaf discs

2.2. Chemicals

- Agar
- Malt extract
- Yeast extract
- KH_2PO_4
- $MgSO_4 \cdot 7H_2O$
- NaCl
- K_2HPO_4
- KNO_3
- KCl
- $(NH_4)_2SO_4$
- NaOH
- 3,5 Dinitrosalicylic acid (DNS)
- Na K tartrate (Rochelle salt)
- Na metabisulphite
- Phenol (melt at 50 °C)
- Deionized water
- Glycine
- Glucose
- Xylose
- Mannose
- Sodium citrate buffer (1 M, pH 4.5)
- Carboxymethylcellulose (2 % w:v in 50 mM sodium citrate buffer)
- *p*-Nitrophenyl-β-glucoside
- *p*-Nitrophenyl-β-xyloside
- Birchwood xylan (1% w:v 50 mM sodium citrate buffer)
- Mannan (1% w:v 50 mM sodium citrate buffer)
- Ivory nut mannan (pure mannan)
- Konjac mannan (glucomannan)
- Softwood galactoglucomannan
- Locust bean mannan or guar gum (galactomannan)
- *p*-Nitrophenyl-β-mannoside

2.3. Solutions

- Mineral solution for fungal growth: 10 mM KNO_3, 2.5 mM KH_2PO_4, 2.5 mM K_2HPO_4, 3 mN NaCl, 1 mM $MgSO_4$; adjust to pH 7 before autoclaving.

- Solution 1: Dinitrosalicylic acid (DNS) reagent:
  - o Measure 801.0 ml of deionized water.
  - o Dissolve 11.2 g NaOH and 6.0 g DNS in about 400 ml of the water in a 1000 ml container.
  - o Use a powder funnel to add 173.2 g Na K tartrate and 4.7 g Na metabisulphite.
  - o Use the remaining water to wash all reagents into the 1000 ml container.
  - o Add 4.3 g phenol melted at 50 °C and stir to dissolve.
- Solution 2: 0.4 M glycine buffer:
  - o Dissolve 60 g of glycine in 1500 ml deionized water.
  - o Add 50% (v:v) NaOH solution until the pH is 10.8.
  - o Dilute to exactly 2000 ml.

3. EXPERIMENTAL PROCEDURES

3.1. Fungal Growth in Liquid Culture

1. Dispense 150 ml of mineral solution to 500-ml Erlenmeyer flask.
2. Add 1—2 g sterile leaf discs to flask.
3. Inoculate aseptically with a 5-mm plug from a 7—14 day old culture.
4. Grow isolate as shake flask culture (ca. 140 rpm) at 15—20 °C for 7—21 days.
5. Transfer content of flask to centrifuge tube and recover culture supernatant by centrifugation (20000 g) for 10 min at 4 °C.

3.2. Endohydrolase Activity

1. Select appropriate standard (e.g., glucose) and substrate (e.g., carboxymethylcellulose).
2. Make up standards stock solution (10 mM of glucose).
3. Dilute standard stock to give dilution standards (2, 4, 6, 8 and 10 µM).
4. Dispense 1.5 ml of each standard (glucose) or substrate (e.g., carboxymethylcellulose) solution in separate 15-ml test tubes.
5. Condition substrate in water bath at 50 °C for at least 5 min (steps 5 – 8 are not necessary for standard solutions).
6. Add 0.5 ml of culture filtrate, or buffer (blank) to each test tube containing substrate.
7. Vortex tube and return to water bath.
8. Incubate for exactly 5 min.
9. Stop reaction by adding 3.0 ml DNS reagent. Add the same amount to tubes with standards.
10. Vortex tube and place directly in boiling water bath for exactly 5 min.
11. Prepare enzyme blank by adding 3.0 ml of DNS to a tube containing substrate, then add 0.5 ml of culture filtrate and place in the boiling water bath for exactly 5 min, then add 3 ml of DNS.

12. Cool tube down in water bath (20 °C).
13. Zero spectrophotometer with reaction blank at 540 nm.
14. Read absorbance of sample at room temperature at 540 nm.
15. Generate linear calibration curve by plotting glucose concentrations in tubes with standard (μmol ml^{-1}) versus absorbance at 540 by forcing line through zero. Obtain slope, y-intercept, and r^2 value, which should be >0.98.
16. Determine net absorbance in substrate tubes by subtracting appropriate culture filtrate blanks from hydrolysis samples (averaged value).
17. Determine sugar concentration (μmol ml^{-1}) liberated by hydrolysis with culture filtrate unknown using line equation (y = mx + b).
18. Calculate enzyme activity, determined as nkat ml^{-1} culture filtrate, by the following equation:

$$\frac{nkat}{ml} = \frac{\mu mol \cdot 1000}{ml \; filtrate \cdot 300 \; s} \tag{32.1}$$

3.3. Filter Paper Activity (Total Cellulase Activity)

1. Prepare glucose standards (6.7, 5.0, 3.3 and 2.0 mg ml^{-1}).
2. Place 50 mg of Whatman No. 1 (1 cm × 6 cm) filter paper into 25-ml test tube.
3. Add 1 ml of 50 mM sodium citrate buffer (pH 4.5).
4. Condition substrate in water bath at 50 °C for at least 5 min.
5. Add 0.5 ml of culture filtrate, sugar stock solutions and blank (buffer) to individual test tubes and mix.
6. Incubate at 50 °C for exactly 60 min.
7. Terminate reaction by adding 3.0 ml of DNS reagent.
8. Vortex tube and place in boiling water bath for exactly 5 min.
9. Cool tube down in water bath (20 °C).
10. Add 10 ml of deionized water.
11. Zero spectrophotometer with buffer blank at 540 nm.
12. Determine absorbance of sample at 540 nm.
13. Generate standard linear curve by plotting absolute amount of sugar (mg 0.5 ml^{-1}) versus absorbance at 540 nm by forcing line through zero. Obtain slope, y-intercept, and r^2 value, which should be >0.98.
14. Determine net absorbance by subtracting culture filtrate blanks from absorbance of hydrolysis samples.
15. Determine concentration of glucose liberated during the reaction with culture filtrate unknown using line equation (y = mx + b); mg Unknown = (net absorbance – y-intercept)/slope
16. Calculate total cellulase activity, determined as Units ml^{-1} of culture filtrate, by the following equation:

$$\frac{Units}{ml} = \frac{mg\ Unknown \cdot 0.5 \cdot 60}{0.18} \qquad (32.1)$$

3.4. ß-Glucosidase Activity

1. Add 1 ml of 5 mM *p*-nitrophenyl-β-glucoside in 50 mM sodium acetate buffer (pH 4.8) to a 10-ml test tube.
2. Add 1.8 ml of 100 mM acetate buffer.
3. Condition substrate in water bath at 50 °C for at least 5 min.
4. To the substrate add 200 µl culture filtrate, and vortex vigorously.
5. Place in water bath at 50 °C for exactly 30 min.
6. Terminate reaction by adding 4 ml 0.4 M glycine buffer.
7. Cool tube down in water bath (20 °C).
8. Use blank of 200 µl fresh culture media to zero spectrophotometer at 430 nm.
9. Determine absorbance of sample at 430 nm.
10. Generate standard linear curve by plotting sugar concentration (μmol ml^{-1}) versus absorbance at 430 nm (correct for dilution of standard concentrations) by forcing line through zero. Obtain slope, y-intercept and r^2 value, which should be >0.99.
11. Since the unknown assays are generally done in duplicate, average the duplicates.
12. Determine net absorbance by subtracting appropriate culture filtrate blanks from hydrolysis samples (averaged value).
13. Determine sugar concentration (μmol ml^{-1}) liberated by hydrolysis with culture filtrate unknown using line equation (y = mx + b).
14. μmol ml^{-1} Unknown = (net absorbance – y-intercept)/slope.
15. Use the following equation to calculate enzyme activity as nkat ml^{-1} of culture filtrate, where 1 nkat is the activity that releases 1 nmol of *p*-nitrophenol equivalent per second during the assay.

$$\frac{nkat}{ml} = \frac{\mu mol \cdot 1000}{ml\ Unknown \cdot 1800\ s} \qquad (32.3)$$

4. FINAL REMARKS

Instead of culture filtrates, extracts from stream-exposed leaves can be used. However, this generally requires preliminary tests to ensure that there is measurable activity.

Activities of hydrolytic enzymes have traditionally been measured at 50 °C, even though this is often far higher than the temperature experienced by fungi in the field. The enzymes are thermostable, and higher incubation temperature allows much briefer incubation periods. The assay measures enzymatic potential; if actual release of sugars under natural conditions is of interest, incubation at ambient stream temperatures is preferable. This may require much longer incubation periods, and precautions will have to be taken to prevent bacterial contamination.

In the endocellulase assay, each set of assays has a reagent blank and a set of standards where buffer and standards are added to the reaction instead of culture filtrate. Additionally, each assay has an enzyme blank where DNS is added to the substrate before the enzyme so that enzyme activity is prevented and the reducing sugars in the culture filtrate can be determined.

Enzyme activity is generally expressed in nkat ml^{-1} of culture filtrate, where 1 nkat is the activity that releases 1 nmol of product (i.e. reducing sugar or *p*-nitrophenol equivalent) per second during the assay. The international unit (IU) is also often used. It represents the release of 1 μμmol of product per minute.

1 katal = 1 mol s^{-1}

1 nkat = 1 nmol s^{-1}

1 IU = 1 μmol min^{-1} = 16.67 nkat

5. REFERENCES

Bailey, M.J., Biely, P. & Poutanen, K. (1992). Interlaboratory testing of methods for assay of xylanase activity. *Journal of Biotechnology*, 23, 257-270.

Fengel, D. & Wegener, G. (1983). *Wood: Chemistry, Ultrastructure, Reactions*. Walter de Gruyter. New York.

Grabber, J.H., Ralph, J. & Hatfield, R.D. (2000). Cross-linking of maize walls by ferulate dimerization and incorporation into lignin. *Journal of Agriculture and Food Chemistry*, 48, 6106-6113.

Ghose, T.K. (1987). Measurement of cellulase activities. *Pure and Applied Chemistry*, 59, 257–268.

Gübitz, G.M., Hayn, M., Sommerauer, M. & Steiner, W. (1996). Mannan-degrading enzymes from *Sclerotium rolfsii*: characterisation and synergism of two endo β-mannanases and a β-mannosidase. *Bioresource Technology*, 58, 127-135.

Gübitz, G.M., Mansfield, S.D., Böhm, D. & Saddler, J.N. (1998). Effect of endoglucanases and hemicellulases in magnetic and flotation deinking of xerographic and laser-printed papers. *Journal of Biotechnology*, 65, 209-215.

Hon, D.N.-S. & Shiraishi, N. (1991). *Wood and Cellulosic Chemistry*. Marcel Dekker. New York.

Mansfield, S.D., Saddler, J.N. & Gübitz, G.M. (1998). Characterization of endoglucanases from the brown rot fungi *Gloeophyllum sepiarium* and *Gloeophyllum trabeum*. *Enzyme and Microbial Technology*, 23, 133-140.

Mansfield, S.D., Mooney, C. & Saddler, J.N. (1999). Substrate and enzyme characteristics that limit cellulose hydrolysis. *Biotechnology Progress*, 15, 804-816.

Nidetzky, B., Hayn, M., Macarron, R. & Steiner, W. (1993). Synergism of *Trichoderma reesei* cellulases while degrading different celluloses. *Biotechnology Letters*, 15, 71-76.

Sjöström, E. (1993). *Wood Chemistry: Fundamentals and Applications*, 2nd ed. Academic Press. London.

Tenkanen, M., Puls, J., Rättö, M. & Viikari, L. (1993). Enzymatic deacetylation of galactoglucmannans. *Applied Microbiology and Biotechnology*, 39,159-165.

van Tilbeurgh, H., Claeyssens, M. & de Bruyne, C.K. (1982). The use of 4-methylumbelliferyl and other chromophoric glycosides in the study of cellulolytic enzymes. *FEMS Letters*, 149,152-156

CHAPTER 33

CELLULASES

MARTIN ZIMMER

Zoologisches Institut: Limnologie, Christian-Albrechts-Universität Kiel, Olshausenstr. 40, 24098 Kiel, Germany.

1. INTRODUCTION

Cellulose is an unbranched chain of several thousand D-glucose monomers formed by β-1,4-glycosidic bonds. *In situ*, numerous parallel poly-glucose chains form insoluble complexes of crystalline cellulose fibres through hydrogen bonds and van-der-Waals forces. These fibres, together with the surrounding lignin matrix, form "lignocellulose", the major component of plant litter, which effectively resists degradation (e.g., Royer & Minshall 2001). Although cellulase activity as a measure of cellulose degradation may be an effective indicator for litter decomposition in terrestrial systems (Savoie & Gourbière 1989; Skambracks & Zimmer, 1998), the correlation between litter mass loss and cellulase activity seems limited in some aquatic systems (Suberkropp & Jenkins 1995).

Native crystalline cellulose is degraded to its glucose units through the synergistic action of several enzyme classes. In fungi (Ljungdahl & Eriksson, 1985) and flagellate gut symbionts of termites (Yamin & Trager 1979), endo-β-1,4-glucanases (endocellulase, C_X-cellulase; EC 3.2.1.4) cleave inner β-1,4-glycosidic bonds and thus generate oligosaccharides. Cellobiohydrolases (exo-β-1,4-glucanase, exocellulase, C1-cellulase; EC 3.2.1.91) split off cellobiose, a glucose dimer, from the non-reducing end of the oligosaccharide chain (Wood & Garcia-Campayo 1990). While amorphous and soluble cellulose may be degraded through the action of endocellulases alone, the degradation of crystalline cellulose requires the activity of an exocellulase (Wood & Garcia-Campayo 1990), at least in the case of common extracellular fungal cellulases. Cellobiase (β-glucosidase, EC 3.2.1.21) cleaves cellobiose into two glucose moieties (Ljungdahl & Eriksson 1985). Possibly, glucohydrolase (EC 3.2.1.74) is also involved in cellulose degradation as a component of the exoglucanase (Goyal et al. 1991), splitting off glucose – instead of cellobiose – from the non-reducing end of poly- and oligosaccharides. Alternatively,

M.A.S. Graça, F. Bärlocher & M.O. Gessner (eds.), Methods to Study Litter Decomposition: A Practical Guide, 249 – 254.

cellobiose can be degraded oxidatively through the action of a cellobiose-quinone-oxidoreductase (EC 1.1.5.1; Evans et al. 1991), which also reduces quinones and phenoxy radicals generated during phenol oxidation, and converts cellobiose to cellobionic acid (Ljungdahl & Eriksson 1985).

In aerobic bacteria, cellulose hydrolysis has been attributed to two types of enzymes adhering to the cell wall and acting like the fungal endocellulases and cellobiases (Ljungdahl & Eriksson 1985; Wood & Garcia-Campayo 1990). Exocellulases have been found in only few bacteria (Rapp & Beermann, 1991). The cleavage of crystalline cellulose by organisms lacking exocellulase appears to be due to an intra-molecular synergism of bacterial endocellulases (Din et al. 1994). Cellulolysis by anaerobic bacteria results from the action of "cellulosomes" (Leschine 1995), which are multi-protein complexes containing endo- or exocellulases as well as xylanases (EC 3.2.1.37; see Chapter 32) and several other proteins with structural or substrate-binding functions (Wood & Garcia-Campayo 1990).

Endogenous cellulases have also been found in both terrestrial (Slaytor 1992) and aquatic detritivores (Wang et al. 2003; Zimmer & Bartholmé 2003). They are possibly similar to those of aerobic bacteria (Rouland et al. 1988). These cellulases in invertebrates can degrade cellulose without microbial assistance, but they appear to be less effective than microbial cellulases (Slaytor 1992). The acquisition of fungal cellulases through the food may be crucial to supplement critical, non-endogenous cellulase components (Martin 1984).

Endoglucanase activity can be assayed with a water-soluble cellulose derivate, such as carboxymethylcellulose (CMC) or phosphoric acid-swollen cellulose (Chapter 34), but exocellulase activity can only be estimated by measuring the release of glucose from crystalline cellulose such as Avicel, filter paper or cotton. However, the common method of determining the release of reducing groups from cellulose (Ghose 1987; see Chapter 32) is not specific to glucose and is prone to interference by a number of substances in environmental samples (Skambracks & Zimmer 1998). The approach described here uses a more specific, enzyme-based technique to quantify the amount of glucose released from crystalline cellulose by the combined activities of several enzymes (Skambracks & Zimmer 1998).

2. EQUIPMENT AND MATERIALS

2.1. Equipment

- Homogenizer (e.g., electronic disperser or mortar and pestle for leaf litter; rotation grinder, or ultrasonic disintegrator for gut and faeces samples)
- Incubation tubes; e.g., glass tubes with screw caps (15—20 ml) for leaf litter; plastic reaction tubes (1.5 ml) for gut and faeces samples
- Analytical balance
- Shaker
- Centrifuge

- Micropipettes (100—1000 µl; 10—100 µl)
- Plastic cuvettes
- Spectrophotometer

2.2. Material

- Field-collected leaf litter
- Dissected guts of detritivores having fed on leaf litter (gut epithelium best removed)
- Faeces of detritivores having fed on leaf litter

2.3. Chemicals

- α-Cellulose
- KH_2PO_4
- Na_2HPO_4
- Citric acid monohydrate (citrate)
- NaN_3
- Commercially available kit for the determination of glucose and fructose in food (e.g., R-Biopharm, Darmstadt, Germany); if the required solutions are prepared in the laboratory:
  - Tri-ethanolamine-HCl
  - $MgSO_4 \cdot 7H_2O$
  - NaOH, 5 mol l^{-1}
  - NADP-Na_2
  - ATP-Na_2H_2
  - $NaHCO_3$
- Hexokinase (e.g., Sigma)
- Glucose-6-phosphate-dehydrogenase (e.g., Sigma)
- Double-distilled water (H_2O)

2.4. Solutions

- Solution 1 (contained in kit): 0.75 M tri-ethanolamine buffer (pH 7.6), including 10 mM NADP, 80 mM ATP
- Solution 2 (contained in kit): 2 mg ml^{-1} hexokinase and 1 mg ml^{-1} glucose-6-phosphate-dehydrogenase
- Solution 3: 0.05 M K-Na-phosphate buffer: 415 ml 0.1 M KH_2PO_4 + 85 ml 0.1 M Na_2HPO_4 + 500 ml double-distilled water, pH 6.2; if prepared accurately, pH does not need to be adjusted.
- Solution 4: Citrate-phosphate buffer: 400 ml 0.1 M citrate (citric acid monohydrate) + 600 ml 0.2 M Na_2HPO_4, pH 5.8; if prepared accurately, pH does not need not be adjusted.

- Solution 5: Citrate-phosphate buffer, pH 5.8, with 0.05% NaN$_3$

3. EXPERIMENTAL PROCEDURES

3.1. Extraction of Microbial Enzymes

1. Weigh samples of leaf litter (corresponding to 50—100 mg dry mass), dissected guts (5—10 mg), or faeces (5—10 mg).
2. Determine dry mass:fresh mass-ratios to estimate dry mass of samples from fresh mass.
3. The appropriate method of enzyme extraction depends on the source of enzymes; in case of extracellular enzymes, thoroughly chopping up sample material with a homogenizer is sufficient for accurate measurement of enzyme activity; with cell-bound enzymes, additional sonication is recommended.
4. Homogenize litter samples with the appropriate method in 10 ml of 0.05 M phosphate buffer, or gut or faeces samples in 1 ml of 0.05 M phosphate buffer; although cellulases are quite stable, place samples on ice during homogenization to avoid thermal denaturation of enzymes.
5. Homogenates may be stored frozen (–20 °C) until used for assays.
6. Centrifuge suspensions (5 min; ca. 10000 g, depending on the available centrifuge and reaction tubes; 4 °C).
7. Use supernatants for extracellular cellulase activity; pellets can be used for estimating cellular enzyme activities (such as dehydrogenases).

3.2. Determination of Cellulase Activity

1. Add 20 mg α-cellulose to 200 μl-aliquots of supernatant solution.
2. Add 200 μl citrate-phosphate buffer with 0.05 % NaN$_3$.
3. Incubate on a shaker (18—24 h, 20 °C).
4. Centrifuge for 5 min (ca 10000 g, depending on the available centrifuge and reaction tubes; 4 °C).
5. Add 50 μl of the supernatant to 450 μl of Solution 1.
6. Add 1000 μl of double-distilled water.
7. Measure absorbance (A_0) at 340 nm.
8. Add 10 μl of Solution 2.
9. Incubate for 30 min at room temperature.
10. Measure absorbance (A_{30}) at 340 nm.
11. Calculate ΔA as A_{30}–A_0.

12. Calculate glucose concentration, c, of the sample in mg ml^{-1}:

$$c = \frac{test\ volume \cdot molecular\ weight\ of\ glucose}{extinction\ coefficient\ of\ NADPH \cdot cuvette\ length \cdot sample\ volume \cdot 1000}$$

$$= \frac{1.51 \cdot 180.16}{6.3 \cdot 1.0 \cdot 0.05 \cdot 1000} \left(\frac{ml \cdot \dfrac{g}{mol}}{\dfrac{l}{mmol \cdot cm} \cdot cm \cdot ml \cdot 1000} \right) \tag{33.1}$$

13. Run controls without adding α-cellulose to the incubation, and subtract c values from those of samples before calculating cellulase activity.

14. Calculate cellulase activity (μg glucose mg^{-1} h^{-1}):

$$cellulase\ activity = \frac{c \cdot 0.4 \cdot dilution\ factor \cdot 1000}{sample\ dry\ mass \cdot incubation\ time} \left(\frac{\dfrac{mg}{ml} \cdot ml \cdot 1000}{mg \cdot h} \right) \tag{33.2}$$

where the dilution factor = 50 for litter and 5 for guts and faeces,
sample dry mass = 50—100 mg for litter and 5—10 mg for guts and faeces, and
incubation time = 18—24 h.

4. FINAL REMARKS

Preparation of Solution 1 in the laboratory, if no kit is used:
- Solution 1a: dissolve 7.0 g tri-ethanolamine-HCl and 0.125 g MgSO$_4$ · 7H$_2$O in 40 ml H$_2$O; add ca. 2 ml NaOH (5 M) to adjust pH to 7.6; add H$_2$O to 50 ml.
- Solution 1b: dissolve 25 mg NADP-Na$_2$ in 2.5 ml H$_2$O.
- Solution 1c: dissolve 125 mg ATP-Na$_2$H$_2$ and 124 mg NaHCO$_3$ in 2.5 ml H$_2$O
- Mix 50 ml of Solution 1a with 2.5 ml of Solution 1b and 2.5 ml of Solution 1c; store at 4 °C for up to 4 weeks.

5. REFERENCES

Din, N., Damude, H.G., Gilkes, N.R., Miller, R.C., Warren, A.J. & Kilburn, D.G. (1994). C1-Cx revisited: intramolecular synergism in a cellulase. *Proceedings of the National Academy of Sciences, USA*, 91, 11383-11387.

Evans, C.S., Gallagher, I.M., Atkey, P. T. & Wood, D.A. (1991). Localisation of degradative enzymes in white-rot decay of lignocellulose. *Biodegradation*, 2, 93-106.

Ghose, T.K. (1987). Measurement of cellulase activities. *Pure and Applied Chemistry*, 59, 257–268.

Goyal, A., Ghosh, B. & Eveleigh D.E. (1991). Characteristics of fungal cellulases. *Bioresource Technology*, 36, 37-50.

Leschine, S.B. (1995). Cellulose degradation in anaerobic environments. *Annual Review of Microbiology*, 49, 399-426.

Ljungdahl, L.G. & Eriksson, K.-E. (1985). Ecology of microbial cellulose degradation. *Advances in Microbial Ecology*, 8, 237-299.

Martin, M.M. (1984). The role of ingested enzymes in the digestive processes of insects. In J.M. Anderson, A.D.M. Rayner, & D.W.H. Walton (eds.), *Invertebrate-Microbial Interactions* (pp. 155-172). Cambridge University Press. Cambridge.

Rapp, P. & Beermann, A. (1991). Bacterial cellulases. In: C.H. Haigler, & P.J. Weimer (eds.), *Biosynthesis and biodegradation of cellulose* (pp. 269-276). New York: Marcel Dekker.

Rouland, C., Civas, A., Renoux, J. & Petek, F. (1988). Purification and properties of cellulases from the termite *Macrotermes mülleri* (Termitidae, Macrotermitinae) and its symbiotic fungus *Termitomyces* sp. *Comparative Biochemistry and Physiology*, 91B, 449-458.

Royer, T.V. & Minshall, G.W. (2001). Effects of nutrient enrichment and leaf quality on the breakdown of leaves in a hardwater stream. *Freshwater Biology*, 46, 603-610.

Savoie, J.-M. & Gourbière, F. (1989). Decomposition of cellulose by the species of fungal succession degrading *Abies alba* needles. *FEMS Microbiology, Ecology*, 62, 307-314.

Skambracks, D. & Zimmer, M. (1998). Combined methods for the determination of microbial activity in leaf litter. *European Journal of Soil Biology*, 34, 105-110.

Slaytor, M. (1992). Cellulose digestion in termites and cockroaches: what role do symbionts play? *Comparative Biochemistry and Physiology*, 103B, 775-784.

Suberkropp, K. & Jenkins, C.C. (1995). The influence of water chemistry on the enzymatic degradation of leaves in streams. *Freshwater Biology*, 33, 245-253.

Wood, T. M. & Garcia-Campayo, V. (1990). Enzymology of cellulose degradation. *Biodegradation*, 1, 147-161.

Yamin, M.A. & Trager, W. (1979). Cellulolytic activity of an axenically-cultivated termite flagellate, *Trichomitopsis termopsidis. Journal of General Microbiology*, 13, 417-420.

Wang, J., Ding, M., Li, Y.-H., Chen, Q.-X., Xu, G.-J. & Zhao F.-K. (2003). Isolation of a multi-functional endogenous cellulase gene from mollusc, *Ampullaria crossean. Acta Biochimica et Biophysica Sinica*, 35, 941-946.

Zimmer, M. & Bartholmé, S. (2003). Bacterial endosymbionts in *Asellus aquaticus* (Isopoda) and *Gammarus pulex* (Amphipoda), and their contribution to digestion. *Limnology and Oceanography*, 48, 2208-2213.

CHAPTER 34

VISCOSIMETRIC DETERMINATION OF ENDOCELLULASE ACTIVITY

BJÖRN HENDEL & JÜRGEN MARXSEN

Limnologische Fluss-Station des Max-Planck-Instituts für Limnologie, D-36110 Schlitz, Germany.

1. INTRODUCTION

Cellulose is an important carbon source in aquatic ecosystems receiving or producing organic matter of vascular plant origin. Heterotrophic microorganisms such as fungi and bacteria cleave the high-molecular weight compounds into smaller fragments, which they incorporate into their own biomass, thus making the carbon available to members of the food webs that lack the capability of hydrolyzing native cellulose.

Enzymatic degradation of cellulose may involve up to three types of extracellular enzymes working synergistically to transform the polymeric cellulose molecule to glucose monomers (Robson & Chambliss 1989, Gilbert & Hazlewood 1993; Chapters 32 and 33): (1) Endocellulases (endo-β-1,4-glucanases) cleave internal β-1,4-glycosidic bonds randomly within the native chain of cellulose; (2) exocellulases (exo-β-1,4-glucanases, mainly cellobiohydrolases) release cellobiose (or glucose) from the non-reducing ends of cellulose; and (3) β-glucosidases (cellobiases) hydrolyse cellobiose into two glucose units.

The activity of β-glucosidases and exocellulases can be determined by means of fluorogenic (or chromogenic) model substrates, whereas endocellulase activity can be estimated by monitoring viscosity of a standard cellulose solution. The viscosity declines in parallel to the average molecular weight of the dissolved molecules; in the presence of endocellulase activity, the large cellulose chains are cut into smaller fragments. Naturally occurring cellulose is not soluble in water; for the assay, it is therefore replaced by water soluble carboxymethylcellulose, in which numerous hydroxy groups are substituted by carboxymethyl groups. Carboxymethylcellulose does, however, provide the same β-1,4-glycosidic bonds as natural cellulose that are the target of cellulases (Hulme 1988).

255

M.A.S. Graça, F. Bärlocher & M.O. Gessner (eds.), Methods to Study Litter Decomposition: A Practical Guide, 255–260.
© 2005 *Springer. Printed in The Netherlands.*

The reduction of viscosity is monitored in vertical glass capillary tubes (Micro-Ubbelohde viscometer, e.g. Schott) containing a solution of carboxymethylcellulose and enzymes. The efflux time of the solution in the viscometer is recorded as a measure of viscosity, and standardized enzyme units are calculated from these data. The procedure presented here has been adopted from Almin & Eriksson (1967) and Hendel (1999).

2. EQUIPMENT, CHEMICALS AND SOLUTIONS

2.1 Equipment

- Sharp knife or scalpel
- Homogenizer (e.g. Polytron)
- Homogenization vessels, 100 ml
- Water bath (5 °C)
- Bench centrifuge
- Micro-Ubbelohde viscometer: volume 5 ml, capillary diameter approximately 0.5 mm (Fig. 34.1)
- Viscometer tripod
- Automatic viscosity system or stopwatch with water bath at 25.0 °C. An automatic system (e.g. ViscoSystem AVS 350, Schott Geräte GmbH, Hofheim am Taunus, Germany) replaces manual measurement. It consists of a water bath with controlled temperature. A light barrier measures the time needed for the surface of the solution to fall from mark tm1 to tm2 (Fig. 34.1).
- Gooch crucibles (20 ml, porosity 4 with an approximate pore width of 10—16 μm)
- Pipettes, Eppendorf type or equivalent (e.g. 2000 μl and 5000 μl)

2.2 Chemicals

- Carboxymethylcellulose, substitution grade 0.5—0.7
- Acetic acid 100%, analytical grade
- Sodium hydroxide, analytical grade
- Autoclaved water from sampling site

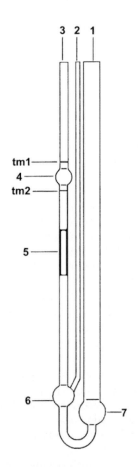

Fig. 34.1: Ubbelohde micro-viscometer. 1: Filling tube, 2: Venting tube, 3: Capillary tube, 4: Measuring sphere, 5: Capillary, 6: rence level vessel, 7: Reservoir, tm1: Upper timing mark, tm2: Lower timing mark

2.3 Solutions

- Acetate buffer: 50 mM (3.0025 g acetic acid 100%). Adjust pH with NaOH to 5.0.
- Carboxymethylcellulose stock solution: 50 g l^{-1} in acetate buffer. Dissolve pellets or powder overnight on a magnetic stirrer. Filter stock solution through a Gooch crucible and freeze immediately in appropriate portions (100—200 ml depending on the number of processed samples per day) at –18 °C. Use this stock solution for all samples. Thaw portions to be used in assay just before the analysis.

3. EXPERIMENTAL PROCEDURES

3.1. Sample Preparation

1. Collect wood, leaves or other type of organic matter and transport to the laboratory in a cooled, insulated container. The fresh mass of samples should be ≥1 g. Process sample as soon as possible, but no later than 6 h after collection.
2. Remove any adhering debris and macroinvertebrates.
3. With a knife or scalpel cut samples in pieces. The final size depends on the homogenizer used; it must be able to homogenize them to a point where fragments are no longer visible.
4. Transfer pre-cut pieces of particulate organic matter (wet weight ≥1 g) into a homogenization vessel containing unfiltered, autoclaved water from the sampling site (e.g. 80 ml for a 100 ml vessel).
5. Place homogenization vessel in a water bath at 5 °C and homogenize samples for 2 min. Prevent sample temperature from rising above 18 °C.
6. Use the homogenate for enzyme assays.

3.2. Enzyme Analysis

1. With a wide-mouth pipette remove 5 ml of the homogenate from each homogenization vessel and transfer to a clean crucible.
2. Filter each sample and collect the filtrate, which contains the enzymes, in a centrifugation tube.
3. Centrifuge for 8 min at ca. 3000 g to separate any remaining solids from the fluid.
4. Mix 1.5 ml of enzyme solution with 5 ml of carboxymethylcellulose stock solution.
5. Immediately transfer 4 ml of the mixture to the filling tube of the viscometer.
6. Measure efflux time of the solution (the time needed to fall from the upper to the lower mark; tm1 and tm2; see Fig. 34.1) five times at 5-min intervals at 25.00±0.05 °C.

3.3. Calculation

Calculate enzyme activity using the following equation (Hulme 1988):

$$A = 1.27 \frac{dx}{dt} \left(\frac{1}{\eta_{sp}} \right)_{t=0}$$

(34.1)

where A = enzyme activity, and
η_{sp} = specific viscosity of the sample solution. This can be calculated as:

$$\eta_{sp} = \frac{t_s}{t_0} - 1$$

(34.2)

where t_s = efflux time of enzyme solution, and
t_0 = efflux time of the acetate buffer.

The derivative dx/dt of equation (1) is calculated from the plot of $1/\eta_{sp}$ against elapsed time. At least the first three, and generally all five, measurements lie on a straight line, whose slope is the required value. The intercept has no meaning.

The result appears as international enzyme units (IEU). One IEU corresponds to the amount of enzyme that catalyzes the hydrolysis of one microequivalent of β-1,4-glucosidic bonds per min at defined conditions of pH and temperature (International Union of Biochemistry 1984).

4. FINAL REMARKS

The use of carboxymethylcellulose with an exactly defined substitution grade is critically important (Eriksson & Hollmark 1969), as is maintenance of the temperature during the assay within an interval of ±0.05 °C around 25 °C. Filtering the homogenized solution can be replaced by centrifugation at 38000 g for 20 min.

Processing of one sample in the viscometer requires at least 40 min. When a single viscometer is available, about 12 samples can be processed during a normal work day.

Best results are obtained with an automatic viscosity system. Such a system minimizes measurement error when determining the time taken by the liquid to fall from tm1 and tm2 (Fig. 34.1), due to automatic light barriers and exact temperature adjustment. Nevertheless, a standard capillary with a defined diameter of approximately 0.5 mm in a water bath with a defined temperature can yield accurate and sufficiently precise values, although the error tends to be significantly larger.

Only a few published values are available for endocellulase activity from particulate organic matter in aquatic systems. Decomposing alder (*Alnus glutinosa* L.) and beech (*Fagus sylvatica* L.) wood in an upland stream had 50—550 IEU g^{-1} ash-free dry mass (AFDM); values for decomposing leaves of the same species ranged from 25—1000 IEU g^{-1} AFDM (Hendel 1999). Dogwood (*Cornus florida*

L.), red maple (*Acer rubrum* L.) and chestnut oak (*Quercus prinus* L.) leaves decomposing in another stream ranged from <1000—9000 IEU g^{-1} AFDM (Linkins et al. 1990).

5. REFERENCES

Almin, K.E. & Eriksson, K.-E. (1967). Enzymic degradation of polymers. I. Viscosimetric method for the determination of enzymic activity. *Biochimica et Biophysica Acta, 139,* 238-247.

Eriksson, K.-E. & Hollmark, B.H. (1969). Kinetic studies of the action of cellulase upon sodium carboxymethyl cellulose. *Archives of Biochemistry and Biophysics, 133,* 233-237.

Gilbert, H.J. & Hazlewood, G.P. (1993). Bacterial cellulases and xylanases. *Journal of General Microbiology, 139,* 187-194.

Hendel, B. (1999). *Der mikrobielle Abbau von Holz und Laub im Breitenbach unter besonderer Berücksichtigung der Bedeutung extrazellulärer Enzyme.* PhD Thesis. University of Gießen, Germany.

Hulme, M.A. (1988). Viscosimetric determination of carboxymethylcellulase activity. In W. A. Wood & S. T. Kellogg (eds.), *Methods in Enzymology Vol. 160, Biomass, Part A, Cellulose and Hemicellulose* (pp. 130-135). Academic Press. New York.

International Union of Biochemistry. (1984). *Enzyme Nomenclature: Recommendations of the Commission on Biochemical Nomenclature.* Academic Press. New York.

Linkins, A.E., Sinsabaugh, R.L., McClaugherty, C.A. & Melillo, J.M. (1990). Comparison of cellulase activity on decomposing leaves in a hardwood forest and woodland stream. *Soil Biology and Biochemistry, 22,* 423-425.

Robson, L.M. & Chambliss, G.H. (1989). Cellulases of bacterial origin. *Enzyme and Microbial Technology, 11,* 626-644.

CHAPTER 35

FLUOROMETRIC DETERMINATION OF THE ACTIVITY OF β-GLUCOSIDASE AND OTHER EXTRACELLULAR HYDROLYTIC ENZYMES

BJÖRN HENDEL & JÜRGEN MARXSEN

Limnologische Fluss-Station des Max-Planck-Instituts für Limnologie, D-36110 Schlitz, Germany.

1. INTRODUCTION

Cellulose is the major structural polysaccharide of vascular plants and cannot be broken down by most animals (Chapter 32). Like other polymers, it is too large to be taken up by microbial cells; it first has to be cleaved into smaller subunits. This occurs via extracellular enzymes, produced mainly by bacteria and fungi. These enzymes are bound to cell surfaces or released into the environment. Thus, cleavage of cellulose and other macromolecular compounds by extracellular enzymes is a crucial initial step in the microbially mediated degradation of leaf litter and wood (Marxsen & Fiebig 1993), and is also essential for the regeneration of inorganic nutrients (Marxsen & Schmidt 1993).

Activities of many hydrolytic extracellular enzymes can be measured precisely by means of fluorogenic model substrates, which are available for a suite of natural compounds (Hoppe 1983). These model substrates consist of a 4-methylumbelliferone (MUF) molecule linked to another compound (e.g. glucose, phosphate, an amino acid) and are cleaved by the enzyme in a way similar to the natural oligomeric or polymeric substances (Fig. 35.1). The MUF released by hydrolysis is fluorescent. Its quantity can be determined fluorometrically and indicates the level of extracellular enzyme activity in a sample.

Free methylumbelliferone exhibits its maximum fluorescence at an excitation wavelength between 355 and 380 nm at a pH above 10. The emission wavelength is between 440 and 460 nm. Humic compounds occurring in many samples from aquatic ecosystems may interfere and shift excitation and emission wavelengths. Consequently, before any measurements, the specific excitation and emission

M.A.S. Graça, F. Bärlocher & M.O. Gessner (eds.), Methods to Study Litter Decomposition: A Practical Guide, 261 – 266.

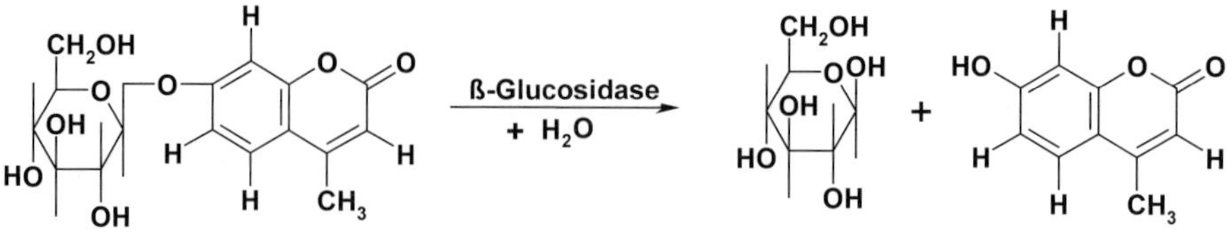

Figure 35.1. Example of a MUF-compound and its cleavage: hydrolysis of non-fluorescent MUF-β-glucoside into glucose and fluorescent MUF

wavelengths for the samples being studied need to be determined. A main advantage of the technique is its high sensitivity, which is several orders of magnitudes above that of methods based on chromogenic substances like pNP(*p*-nitrophenyl)-compounds (Tank et al. 1998, Chapter 32). Short incubation periods (60 min) and incubation at *in situ*-temperatures are therefore possible.

MUF model compounds are available for several natural substances (Table 35.1). The method for β-glucosidase described below can be modified for the detection of a broad range of extracellular enzymes occurring in aquatic systems. In addition to carbohydrate-degrading enzymes like xylosidase, cellobiohydrolase, and chitinase, activities of enzymes degrading other major biopolymers, such as proteins and lipids, and of enzymes involved in nutrient remineralization, such as phosphatases, can be quantified (Hoppe 1993, Marxsen et al. 1998). The procedure presented here has been adopted from Hendel & Marxsen (2000).

Table 35.1. Model substrates with 4-methylumbelliferone (MUF), the corresponding natural substrates and the enzymes cleaving these molecules

Model substrate	Natural substrates	Enzymes
MUF-α-glucoside	maltose, starch	α-glucosidases
MUF-β-glucoside	cellobiose, cellulose	β-glucosidases
MUF-β-xyloside	hemicellulose, xylanes	β-xylosidases
Leucine-MCA	peptides, proteins	peptidases
MUF-laurate	lipids	lipases
MUF-phosphate	polyphosphates	phosphatases

2. EQUIPMENT, CHEMICALS AND SOLUTIONS

2.1. Equipment and Materials

- Sharp knife or scalpel
- Homogenizing device (e.g. Polytron)
- Homogenization vessels (100 ml)
- Water baths (5 and 10 °C)
- Erlenmeyer flasks (25 ml)
- Fluorometer with fluorescence quartz glass cuvettes
- Pipettes

2.2. Chemicals

- Methylumbelliferone (MUF), analytical grade
- Methylumbelliferyl-β-D-glucopyranoside (MUF-glc), analytical grade
- 2-Methoxyethanol (methyl cellosolve, MCS), analytical grade
- Glycine, analytical grade
- Ammonia (25%), analytical grade
- Sodium hydroxide pellets, analytical grade
- Autoclaved water from sampling site

2.3. Solutions

- MUF stock solution, 300 µmol l^{-1}
- MUF-glc stock solution, 5 mmol l^{-1}
- Ammonium glycine buffer (pH 10.5): dissolve 3.75 g glycine in 14.8 ml ammonia (25%), make up to 1000 ml with deionized water and adjust pH with sodium hydroxide solution to 10.5.

3. EXPERIMENTAL PROCEDURES

3.1. Sample Preparation

1. Collect wood, leaves or other organic matter and transport to the laboratory in a cooled, insulated container. The fresh mass of samples should ≥1 g. Process sample as soon as possible, but no later than 6 h after collection.
2. Remove any adhering debris and macroinvertebrates. With a knife or scalpel cut samples in pieces. The final size depends on the homogenizer used; it must be able to homogenize them to a point where fragments are no longer visible.
3. Transfer representative pieces of organic matter (wet weight ≥1 g) into a homogenization vessel containing unfiltered, autoclaved water from the sampling site (e.g. 80 ml for a 100 ml vessel).
4. Place homogenization vessel in a water bath at about 5 °C and homogenize samples for 2 min. Prevent sample temperatures from rising above 18 °C.
5. Use the homogenate for enzyme assays.

3.2. Enzyme Analysis

1. Determine the wavelength of maximum excitation and emission of a water sample with the fluorimeter.

3.2.1. Incubation
1. Prepare fresh stock solutions each time. Add 1% (v:v) MCS to enhance solubility of MUF substrates.

2. Prepare the incubation vessels (Erlenmeyer flasks). Analyse at least 4 sample replicates and an appropriate number of calibration standards (≥3) and blanks (≥2).
3. Transfer 200 µl of leaf or wood homogenate into each Erlenmeyer flask (samples, calibration standards and blanks).
4. For samples, add 1 ml of MUF-glc solution to the homogenate; for calibration standards, add 660 µl of MUF solution. Make up the final volume in each flask to 20 ml.
5. Place each flask in a shaking water bath at 10 °C. Shaking reduces cell adhesion to the flask walls.

3.2.2. Measurement of Enzyme Activity
1. After 5 min, remove 4 ml from each flask and transfer to a centrifuge tube.
2. Add 400 µl of ammonium glycine buffer to each flask to raise pH. This stops enzyme activity and converts MUF to its anionic form, which amplifies the fluorescence.
3. Shake flask gently.
4. Centrifuge for 1 min at ~3000 g.
5. Calibrate the fluorimeter, measure fluorescence of the sample and correct for naturally occurring fluorescence (blanks).
6. Repeat the procedure for measuring enzyme activity after a further 60 min.

3.2.3. Calculation of Enzyme Activity
1. Calculate the enzyme activity in the samples based on the difference between MUF concentrations after 5 and 65 min. Assume a linear increase.

4. FINAL REMARKS

The described method is adjusted for application with a single substrate concentration, which is usually in the saturation range of the enzyme (here β-glucosidase). If it is necessary to determine complete enzyme kinetics, concentrations ranging from one order of magnitude below K_m to one order of magnitude above should be investigated. If substrates other than MUF-glc are used, we recommend first determining the substrate concentration required for saturation.

After appropriate adjustment, the procedure can be used to quantify extracellular enzyme activity in a variety of sample types. In addition to leaves and wood (Hendel & Marxsen 2000), it has been applied successfully to sediments (e.g. Marxsen & Fiebig 1993, Marxsen et al. 1998), biofilms (e.g. Freeman et al. 1990, Romaní 2000) and water (Hoppe 1983, 1993), including water from springs or groundwater with low enzyme activity (Hendel & Marxsen 1997).

Activities of polysaccharide-degrading enzymes associated with particulate organic matter in aquatic systems have been determined with fluorogenic model substrates by Hendel (1999) and Hendel & Marxsen (2000). Activities of β-glucosidase, cellobiohydrolase, β-xylosidase were, respectively, about 1—2000, 1—150, or 4—300 µmol g^{-1} AFDM h^{-1} on decomposing wood of alder (*Alnus glutinosa*

L.) and beech (*Fagus sylvatica* L.). Activities on decomposing leaves of the same tree species were about 100—3000, 1—500, 20—300 µmol g^{-1} AFDM h^{-1}. Activities in the lower range of, or one order of magnitude below, the values reported above were observed on wood in North American streams (Sinsabaugh et al. 1992, Tank et al. 1998).

5. REFERENCES

Freeman, C., Lock, M.A., Marxsen, J., & Jones, S. (1990). Inhibitory effects of high molecular weight dissolved organic matter upon metabolic processes in biofilms from contrasted rivers and streams. *Freshwater Biology*, 24, 159–166.

Hendel, B. (1999). *Der mikrobielle Abbau von Holz und Laub im Breitenbach unter besonderer Berücksichtigung der Bedeutung extrazellulärer Enzyme*. PhD. Thesis. University of Gießen, Germany.

Hendel, B., & Marxsen, J. (1997). Measurement of low level extracellular enzyme activity in natural waters using fluorigenic model substrates. *Acta Hydrochimica et Hydrobiologica*, 25, 253–258.

Hendel, B., & Marxsen, J. (2000). Extracellular enzyme activity associated with degradation of beech wood in a Central European stream. *International Review of Hydrobiology*, 85, 95–105.

Hoppe, H.-G. (1983). Significance of exoenzymatic activities in the ecology of brackish water: measurements by means of methylumbelliferyl-substrates. *Marine Ecology – Progress Series*, 11, 299-308.

Hoppe, H.-G. (1993). Use of fluorogenic model substrates for extracellular enzyme activity (EEA) measurement of bacteria. In P.F. Kemp, B.F. Sherr, E.B. Sherr & J.J. Cole (eds.), *Handbook of Methods in Aquatic Microbial Ecology* (pp. 423-431). Lewis Publishers. Boca Raton.

Marxsen, J., & Fiebig, D.M. (1993). Use of perfused cores for evaluating extracellular enzyme activity in stream-bed sediments. *FEMS Microbiology Ecology*, 13, 1–11.

Marxsen, J., & Schmidt, H.-H. (1993). Extracellular phosphatase activity in sediments of the Breitenbach, a Central European mountain stream. *Hydrobiologia*, 253, 207–216.

Marxsen, J., Tippmann, P., Heininger, H., Preuß, G. & Remde, A. (1998). Enzymaktivität. In: Vereinigung für Allgemeine und Angewandte Mikrobiologie (ed.), *Mikrobiologische Charakterisierung aquatischer Sedimente – Methodensammlung* (pp. 87-114). Oldenbourg. München.

Romaní, A.M. (2000). Characterization of extracellular enzyme kinetics in two Mediterranean streams. *Archiv für Hydrobiologie*, 148, 99–117.

Sinsabaugh, R.L., Antibus, R.K., Linkins, A.E., McClaugherty, C.A., Rayburn, L., Repert, D. & Weiland, T. (1992). Wood decomposition over a first-order watershed: mass loss as a function of lignocellulose activity. *Soil Biology and Biochemistry*, 24, 743–749.

Tank, J.L., Webster, J.R., Benfield, E.F., & Sinsabaugh, R.L. (1998). Effect of leaf litter exclusion on microbial enzyme activity associated with wood biofilms in streams. *Journal of the North American Benthological Society*, 17, 95–103.

CHAPTER 36

PECTIN-DEGRADING ENZYMES: POLYGALACTURONASE AND PECTIN LYASE

KELLER SUBERKROPP

Department of Biological Sciences, Box 870206, University of Alabama, Tuscaloosa, AL 35487, USA.

1. INTRODUCTION

Pectic substances, or pectins, are most abundant in the middle lamellae of primary cell walls, where their main function is to cement plant cells together into tissues. Pectins are the initial polysaccharide substrates encountered by decomposers in nonlignified or weakly lignified plant tissue, and their removal exposes other polysaccharides such as xylans, mannans, and cellulose to microbial degradation (Chamier & Dixon 1982). Pectin-degrading enzymes produced by plant pathogenic fungi have been implicated in the maceration of living plant tissue (Bateman and Basham 1976, Friend 1977). Macerating activity has also been noted in leaf litter exposed to aquatic hyphomycetes, suggesting that this process is important in the decomposition of plant litter in streams (Suberkropp & Klug 1980, Chamier & Dixon 1982, 1983).

Pectin is a polymer of galacturonic acid in which various percentages of the carboxyl residues have been methylated. There are three major classes of enzymes that degrade pectin. These include (1) hydrolytic enzymes, such as polygalacturonases, that hydrolyze the glycosidic bonds between the galacturonic acid residues; (2) enzymes, such as pectin lyase, that cleave the glycosidic bonds between the galacturonic residues by β-elimination; and (3) esterases, such as pectin methyl esterases, that cleave the methyl group from the galacturonic acid residues (Rexová-Benková & Markovič 1976).

Some pectinases cleave the glycosidic bonds between galacturonic acid residues randomly within the polysaccharides (endopectinases) whereas others cleave bonds and release subunits from the ends of the polysaccharides (exopectinases).

M.A.S. Graça, F. Bärlocher & M.O. Gessner (eds.), Methods to Study Litter Decomposition: A Practical Guide, 267 – 272.

Endopectinases cause a reduction in the viscosity of pectin solutions. Consequently, a useful assay for endopectinases involves following decreases in the viscosity of the reaction mixture using viscometers (Chamier & Dixon 1982).

Assays for the hydrolytic enzymes and lyases which depolymerize pectic polysaccharides are presented below. The proposed protocol is based on Suberkropp et al. (1983) and Jenkins & Suberkropp (1995).

2. EQUIPMENT, CHEMICALS AND SOLUTIONS

2.1. Equipment and Material

- Spectrophotometer or colorimeter (550 and 540 nm)
- pH meter
- Magnetic stirrer and hot plate
- Water bath (set at stream temperature)
- Test tubes
- Glassware for preparing and storing solutions
- Adjustable pipette (1 ml)
- Dialysis tubing
- Bench centrifuge (9000 g)
- Analytical balance
- Drying oven
- Muffle furnace

2.2. Chemicals

- Filtered stream water
- Water, distilled or deionized
- Pectin
- Polygalacturonic acid
- Potassium acetate
- Bicine
- Calcium chloride ($CaCl_2 \cdot 2H_2O$)
- Thiobarbituric acid
- Dinitrosalicylic acid
- Thimerasol
- Potassium hydroxide (KOH)
- Hydrochloric acid (HCl)

2.3. Solutions

- Solution 1: 1% pectin – add 1 g pectin very slowly to 80 ml of water with constant stirring. When dissolved, dialyze overnight, centrifuge (9000 g, 20 min), add 1 ml of 1% thimerosal (to inhibit microbial growth), and adjust final volume to 100 ml. Store at 4—5 °C.

- Solution 2: 0.5% polygalacturonic acid – add 0.5 g polygalacturonic acid to 80 ml of water. Adjust pH of solution to 5.0 and add 1 ml of 1% thimerosal. Adjust final volume to 100 ml and store at 4—5 °C.
- Solution 3: 0.2 M bicine with 0.03 M $CaCl_2$ – add 3.26 g bicine and 0.44 g $CaCl_2 \cdot 2H_2O$ to ca. 70 ml water, adjust pH to 8.0 with 0.1 M KOH solution, add 1 ml of 1% thimerosal. Adjust final volume to 100 ml and store at 4—5 °C.
- Solution 4: 0.2 M potassium acetate – add 1.96 g potassium acetate to ca. 70 ml water, adjust pH to 5.0 with 0.1 M HCl; add 1 ml of 1% thimerosal. Adjust final volume to 100 ml and store at 4—5 °C.
- Solution 5: 0.04 M thiobarbituric acid solution – add 0.58 g thiobarbituric acid to water and stir. Adjust final volume to 100 ml (Ayers et al. 1976).
- Solution 6: Dinitrosalicylic acid reagent – dissolve 1 g 3,5-dinitrosalicylic acid in 20 ml of 2 M NaOH, and 50 ml water. Add 30 g NaK tartrate and adjust final volume to 100 ml with water. Store in stoppered bottle to protect from CO_2 (Bernfield 1955).
- Solution 7: Galacturonic acid standards – make galacturonic acid solution (500 $\mu g\ ml^{-1}$) by dissolving 50 mg galacturonic acid in 100 ml water. Freeze samples, store at –20 °C and thaw when needed.

3. EXPERIMENTAL PROCEDURES

3.1. Sample Preparation

1. Collect decomposing leaves and transport to laboratory in an ice-chest.
2. Wash leaves in stream water and cut 4 leaf discs (ca. 12 mm diameter) for each replicate (2 discs for the assay and 2 for boiled controls). Place leaf discs in filtered stream water.
3. Boil the leaf discs to be used for the boiled controls for 10 min.

3.2. Pectin Lyase

1. Pipette 2 ml bicine plus $CaCl_2$ (Solution 3) and 1 ml pectin (Solution 1) into reaction tubes and incubate in a water bath set at ambient temperature or at a fixed temperature, depending on the specific question of the study.
2. Add leaf discs to reaction mixtures at 15 s intervals. Incubate for 180 min.
3. Remove 0.5 ml reaction mixture and add it to 2.5 ml each of 0.04 thiobarbituric acid (Solution 5) and 0.1 M HCl. Add 5 ml water and boil for 30 min. Cool tubes to room temperature and measure absorbance at 550 nm (A_{550}).
4. Rinse leaf discs from each tube with water, dry to constant weight at 40—50 °C, weigh to the nearest 0.1 mg, ash at 500 °C for 4 hours to determine leaf AFDM (see Chapter 1).
5. Because there is no commercially available standard for the product containing the double-bond, express enzyme activity as A_{550} g^{-1} leaf material h^{-1} from

differences between absorbance readings from experimental and boiled control discs.

3.3. Polygalacturonase

1. Pipette 2 ml acetate (Solution 4) and 1 ml polygalacturonic acid (Solution 2) into reaction tubes and incubate in water bath set at ambient temperature or at a fixed temperature, depending on the specific question of the study.
2. Add leaf discs to reaction mixtures at 15 s intervals. Incubate for 180 min.
3. Remove 1.0 ml of the reaction mixture and add it to 1.0 ml of dinitrosalicylic acid reagent (Solution 6), place in boiling water for 5 min, cool in running tap water, add 20 ml water, mix and measure absorbance at 540 nm.
4. Rinse leaf discs from each tube with water, dry at 40—50 °C, weigh, and ash at 500 °C for 4 hours and reweigh to determine leaf AFDM.
5. Determine activity as µg galacturonic acid produced g^{-1} leaf litter h^{-1} by comparing absorbance with standard curves prepared with known concentrations of galacturonic acid (0—500 µg ml^{-1}; Solution 7) and subtracting values obtained from the boiled controls.

4. FINAL REMARKS

The extracellular enzymes considered above are typically stable and active over a wide range of temperatures. If compatible with the goal of the study, enzyme reactions can be carried out at higher temperatures (e.g. 30 °C) to increase the rate of catalysis and the amount of product when enzyme activity is very low or not detectable at ambient temperatures. Polygalacturonases typically have an optimum pH around 5, pectin lyases around 8.

5. REFERENCES

Ayers, W.A., Papavizas, G.C. & Diem, A.F. (1976). Polygalacturonase production by *Rhizoctonia solani*. *Phytopathology,* 67, 1250-1258.

Bateman, D.F. & Basham, H.G. (1976). Degradation of plant cell walls and membranes by microbial enzymes. *Physiological Plant Pathology, Encyclopedia of Plant Physiology* (eds. R. Heitefuss & P.U. Williams), New Series Vol. 4, pp. 316-355. Springer Verlag. Berlin.

Bernfield, P. (1955). Amylases, α and β. *Methods in Enzymology*, 1, 149-150.

Chamier, A.-C. & Dixon, P.A. (1982). Pecinases in leaf degradation by aquatic hyphomycetes: the enzymes and leaf maceration. *Journal of General Microbiology*, 128, 2469-2483.

Chamier, A.-C. & Dixon, P.A. (1983). Effects of calcium-ion concentrations on leaf maceration by *Tetrachaetum elegans*. *Transactions of the British Mycological Society*, 81, 415-418.

Friend, J. (1977). Biochemistry of plant pathogens. *Plant Biochemistry II* (ed. D.H. Northcote), Vol. 13, pp. 141-182. University Park Press, Baltimore.

Jenkins, C.C. & Suberkropp, K. (1995). The influence of water chemistry on the enzymatic degradation of leaves in streams. *Freshwater Biology*, 33, 245-253.

Rexová-Benková, L. & Markovič, O. (1976). Pectic enzymes. *Advances in Carbohydrate Chemistry and Biochemistry*, 33, 323-385.

Suberkropp, K., Arsuffi, T.L. & Anderson, J.P. (1983). Comparison of degradative ability, enzymatic activity, and palatability of aquatic hyphomycetes grown on leaf litter. *Applied and Environmental Microbiology*, 46, 237-244.

Suberkropp, K. & Klug, M.J. (1980). The maceration of deciduous leaf litter by aquatic hyphomycetes. *Canadian Journal of Botany*, 58, 1025-1031.

CHAPTER 37

LIGNIN-DEGRADING ENZYMES: PHENOLOXIDASE AND PEROXIDASE

BJÖRN HENDEL[1], ROBERT L. SINSABAUGH[2] &
JÜRGEN MARXSEN[1]

[1]Limnologische Fluss-Station des Max-Planck-Instituts für Limnologie, D-36110 Schlitz, Germany; [2]Department of Biology, University of New Mexico, 167A Castetter Hall, Albuquerque, NM 87131, USA.

1. INTRODUCTION

Lignin is a principal constituent of vascular plants and, after cellulose, the second most abundant naturally occurring compound. Because of its intimate association with cellulose fibres in plants, lignin is important in regulating the flow of carbon into decomposer food webs and, more generally, through ecosystems. Lignin consists primarily of phenylpropane units which are randomly polymerized into three dimensional macromolecules. Degradation of the lignin molecule is an oxidative process that may extend over long periods. The enzymatic equipment for depolymerizing lignin can be found in fungi and bacteria. Several types of enzymes involved in degradation have been described (Kirk & Farrell 1987). These include monooxygenases (phenoloxidases, laccases), dioxygenases and peroxidases. The quantification of the activity of these enzymes in environmental samples is constrained by the solubility of available substrates and the often strong interference by humic molecules.

Assays of oxidative enzyme activity involve a substrate that serves as an electron donor, generating a product that can be quantified spectrophotometrically (Mason 1948). The procedure described here is the most common and uses L-3,4-dihydroxyphenylalanine (L-DOPA) as the electron-donating substrate for the detection of phenoloxidase activity. L-DOPA is a preferred substrate to be used with environmental samples because it is soluble in water and readily oxidized. One of the products of DOPA oxidation has a red tint and can be quantified by measuring its absorbance at a wavelength of 460 nm. When hydrogen peroxide (H_2O_2) is added to the sample, the activity of peroxidases can also be estimated using the same experimental approach.

M.A.S. Graça, F. Bärlocher & M.O. Gessner (eds.), Methods to Study Litter Decomposition: A Practical Guide, 273 – 278.

The method can be used for all types of organic matter occurring in aquatic environments, and it is fast and accurate (Sinsabaugh & Linkins 1990). However, activity associated with particulate organic matter is more easily detected than activity in the water column. Activity also tends to be higher in humic systems and when associated with fine as opposed to coarse organic particles.

There are several important caveats. In particular, the ecological significance of extracellular oxidative enzyme activity can be difficult to interpret (Münster & De Haan 1998, Sinsabaugh & Foreman 2003), because these enzymes are involved in both the formation and degradation of polyphenols (Stevenson 1994) and, in some organisms, they may play a role in mitigating the potentially inhibitory effects of reactive phenols (Freeman et al. 2001). Furthermore, the fact that L-DOPA is readily oxidized means that different classes of oxidative enzymes may contribute variously to colour formation during assays. As a result, the oxidation kinetics may vary for different types of samples. Another potential problem is that L-DOPA can be oxidized non-enzymatically under some conditions. In our experience, samples containing reduced Mn are especially problematic. Thus, appropriate controls and attention to reaction kinetics are important. During assays, strict adherence to an incubation time of 60 min is critical to protect the light-sensitive L-DOPA from further reaction to melanin (Sinsabaugh & Linkins 1990). The specific procedure presented here has been adopted from Sinsabaugh & Linkins (1990), Hendel (1999) and Hendel & Marxsen (2000).

2. EQUIPMENT, CHEMICALS AND SOLUTIONS

2.1. Equipment

- Sharp knife or scalpel
- Homogenizing device (e.g. Polytron)
- Homogenization vessels (100 ml)
- Water baths (5 and 10 °C)
- Shaking incubator (20 °C)
- Bench centrifuge (3000 g)
- Spectrophotometer (460 nm)
- Pipettes, Eppendorf type or equivalent (e.g. 1000 µl and 2500 µl)

2.2. Chemicals

- Acetic acid (100%), analytical grade
- Sodium hydroxide (NaOH), analytical grade
- L-3,4-Dihydroxyphenylalanine (L-DOPA), analytical grade
- Hydrogen peroxide (H_2O_2, 30%), analytical grade

2.3. Solutions

- Acetate buffer: 50 mM (3.0025 g 100% acetic acid per litre, pH adjusted with NaOH to 5.0).
- L-DOPA stock solution: 5 mM (0.98575 g l^{-1}) in 50 mM acetate buffer. As the solution is unstable, prepare immediately before running assays.
- Hydrogen peroxide solution: 0.3% (v:v).

3. EXPERIMENTAL PROCEDURES

3.1. Sample Preparation

1. Collect wood or leaves and transport to the laboratory in a cooled, insulated container. Fresh mass of the samples should be ≥ 1.0 g. Preferably process collected samples immediately.
2. Remove any adhering debris and macroinvertebrates.
3. Crush samples with a knife.
4. Transfer the crushed wood or leaves into a homogenization vessel containing autoclaved water from the sampling site (e.g. 80 ml for a 100 ml vessel).
5. Place the homogenization vessel in a water bath set at about 5 °C and homogenize sample for at least 2 min. Prevent sample temperature from rising above 18 °C.
6. Use the homogenate for enzyme assays.

3.2. Enzyme Analysis

1. Phenoloxidase: mix 2 ml of homogenate with 2 ml of DOPA stock solution. Run at least 4 analytical replicates. As a control, mix 2 ml of the homogenate with 2 ml of acetate buffer.
2. Peroxidase: mix 2 ml of the homogenate with 2 ml of DOPA stock solution. Add 200 µl of hydrogen peroxide solution. Mix gently. Run at least 4 analytical replicates. As a control, mix 2 ml of the homogenate with 2 ml acetate buffer and 200 µl hydrogen peroxide solution.
3. Incubate for exactly 60 min at 20 °C in a shaking incubator.
4. Centrifuge for 1 min at approximately 3000 g.
5. Transfer supernatant to cuvette (size: 1 cm).
6. Measure absorbance immediately at 460 nm in spectrophotometer.

3.3. Calculation of Enzyme Activity

$$\text{Phenoloxidase: } A_{phenolox.} = Abs_{460} / k \qquad (37.1)$$

$$\text{Peroxidase: } A_{perox.} = Abs_{460} / k - A_{phenolox.} \qquad (37.2)$$

A = enzyme activity in International Enzyme Units (IEU)

Abs_{460} = absorbance at 460 nm

k = extinction coefficient, which is 1.66 mM for DOPA under the conditions of this assay.

4. FINAL REMARKS

The pH optimum of the reaction is about 8. Assays should consequently be run at this pH if the goal is to estimate maximum potential activities. In contrast, to assess activities occurring naturally in the environment (e.g. within leaf litter or soils), acetate buffer at pH 5, as used in the protocol here, is often preferable because 5 is the typical pH of litter and soils. However, in some aquatic environments, such as hardwater streams, the intrinsic pH of litter might deviate from 5 and should hence be determined before choosing a pH for carrying out the enzyme assays.

The assay can be adapted for 96-well microplates (Saiya-Cork et al. 2002, Gallo et al. 2004). In this procedure, 200 μl aliquots of sample homogenate are dispensed into replicate wells. For phenoloxidase, 50 μl of 25 mM DOPA is added to each sample well. Peroxidase assays receive 50 μl of 25 mM DOPA plus 10 μl of 0.3% H_2O_2. Negative control wells for phenoloxidase contain 200 μl of acetate buffer and 50 μl of DOPA solution; blank control wells contain 200 μl of sample suspension and 50 μl of acetate buffer. For peroxidase, negative and blank control wells also receive 10 μl of H_2O_2. There are 16 replicate sample wells for each assay and eight replicate wells for blanks and controls. The microplates are incubated in the dark at 20 °C for up to 18 h. Activity is quantified by measuring absorbance at 460 nm using a microplate spectrophotometer. Results are expressed in units of nmol h^{-1} g^{-1} organic matter using a micromolar extinction coefficient of 7.9.

The few data available to date suggest that oxidative enzyme activities associated with decomposing litter in aquatic ecosystems vary widely. Typical average activities of phenoloxidase associated with decomposing wood in streams are 1—20 μmol h^{-1} g^{-1} organic matter, with a maximum of 40 μmol h^{-1} g^{-1} (Sinsabaugh et al. 1992, Tank et al. 1998, Hendel & Marxsen 2000). Somewhat higher activities have been measured for leaf litter (Hendel 1999), and activities up to about 450 μmol h^{-1} g^{-1} were determined occasionally by Alvarez & Guerrero (2000) for coarse-particulate organic matter in shallow Mediterranean ponds. Typical peroxidase activities on both wood and leaf litter in streams are 1—20 μmol h^{-1} g^{-1} organic matter (Sinsabaugh et al. 1992, Tank et al. 1998, Hendel 1999).

5. REFERENCES

Alvarez, S. & Guerrero, M.C. (2000). Enzymatic activities associated with decomposition of particulate organic matter in two shallow ponds. *Soil Biology and Biochemistry*, 32, 1941-1951.

Freeman, C., Ostle, N. & Kang, H. (2001). An 'enzymatic latch' on a global carbon store. *Nature*, 409, 149.

Gallo, M., Amonette, R., Lauber, C., Sinsabaugh, R.L. & Zak, D.R. (2004). Short-term changes in oxidative enzyme activity and microbial community structure in nitrogen-amended north temperate forest soils. *Microbial Ecology*, In press.

Hendel, B. (1999). *Der mikrobielle Abbau von Holz und Laub im Breitenbach unter besonderer Berücksichtigung der Bedeutung extrazellulärer Enzyme*. PhD. Thesis. University of Gießen. Gießen, Germany.

Hendel, B. & Marxsen, J. (2000). Extracellular enzyme activity associated with degradation of beech wood in a Central European stream. *International Review of Hydrobiology*, 85, 95-105.

Kirk, T.K. & Farrell, R.L. (1987). Enzymatic "combustion": The microbial degradation of lignin. *Annual Review of Microbiology*, 41, 465-505.

Mason, H.S. (1948). The chemistry of melanin. III. Mechanism of the oxidation of dihydroxyphenyl-alanine by tyrosinase. *Journal of Biological Chemistry*, 172, 83-99.

Münster, U. & De Haan, H. (1998). The role of microbial extracellular enzymes in the transformation of dissolved organic matter in humic waters. In D.O. Hessen, & L.J. Tranvik (eds.), *Aquatic Humic Substances* (pp. 199-257). Springer Verlag. Berlin.

Saiya-Cork, K.R., Sinsabaugh, R.L. & Zak, D.R. (2002). Effects of long term nitrogen deposition on extracellular enzyme activity in an *Acer saccharum* forest soil. *Soil Biology and Biochemistry*, 34, 1309-1315.

Sinsabaugh, R.L., Antibus, R.K., Linkins, A.E., McClaugherty, C.A., Rayburn, L., Repert, D. & Weiland, T. (1992). Wood decomposition over a first-order watershed: mass loss as a function of lignocellulose activity. *Soil Biology and Biochemistry*, 24, 743-749.

Sinsabaugh, R.L. & Foreman, C.M. (2003). Integrating DOM metabolism and microbial diversity. In S. Findlay, & R.L. Sinsabaugh (eds.), *Aquatic Ecosystems: The Interactivity of Dissolved Organic Matter* (pp. 426-454). Academic Press. San Diego.

Sinsabaugh, R.L. & Linkins, A.E. (1990). Enzymic and chemical analysis of particulate organic matter from a boreal river. *Freshwater Biology*, 23, 301-309.

Stevens, F.J. (1994). Biochemistry of the formation of humic substances. In F.J. Stevenson (ed.), *Humus Chemistry* (pp. 188-211). John Wiley & Sons. New York.

Tank, J.L., Webster, J.R., Benfield, E.F. & Sinsabaugh, R.L. (1998). Effect of leaf litter exclusion on microbial enzyme activity associated with wood biofilms in streams. *Journal of the North American Benthological Society*, 17, 95–103.

CHAPTER 38

PHENOL OXIDATION

MARTIN ZIMMER

Zoologisches Institut: Limnologie, Christian-Albrechts-Universität Kiel, Olshausenstr. 40,
24118 Kiel, Germany.

1. INTRODUCTION

Lignins are major structural components of plant cell walls and therefore of plant litter. They are complex polymers of a small number of methoxylated phenolic compounds such as coumaryl alcohol, sinapyl alcohol, and coniferyl alcohol (Boerjan et al. 2003). Due to strong C-C linkages and alkyl-aryl ether bonds, lignins effectively resist chemical and enzymatic attack (Hagerman & Butler 1991). Therefore, lignin degradation requires phenol oxidation (cf. Breznak & Brune 1994).

Other important phenolic litter constituents include condensed tannins (Harrison 1971; Savoie & Gourbiére 1989; see Chapter 16). Condensed tannins are regularly-structured polymers of flavan-3-ols and flavan-3,4-diols that are linked through C-C bonds between the monomers (Swain 1979, Hagerman & Butler 1991). As with lignins, the degradation of condensed tannins begins with oxidation. Hydrolyzable tannins, in contrast, are glucose esters of gallic acid or ellagic acid units, and are hence subject to hydrolysis by esterases.

The degradation of the lignin moiety of lignocellulose (see Chapter 37) is strongly dependent on microbial activity (Breznak & Brune 1994). However, not every microbial species involved in decomposition is capable of degrading lignocellulose (Ljungdahl & Eriksson 1985). In contrast to brown- and white-rot fungi, which are primarily terrestrial, the litter-degrading soft-rot and other fungi (mostly Ascomycetes and Deuteromycetes) prominent in aquatic environments are only weakly adapted to degrading lignin (Rabinovich et al. 2004).

Numerous enzymes are involved in phenol oxidation. Laccases (EC 1.10.3.2) have been found in plants and many fungi (Mayer 1987), but only in one bacterium (Faure et al. 1995). Tyrosinases (EC 1.14.18.1) are known from fungi (Wood 1980, Claus & Filip 1990), actinomycetes (Claus & Filip 1990) and plants (Summers & Felton 1994). Both laccases and tyrosinases may be important in wood and leaf-litter decomposition (Wood 1980, Thurston 1994). Catechol oxidase (EC 1.10.3.1) has a

M.A.S. Graça, F. Bärlocher & M.O. Gessner (eds.), Methods to Study Litter Decomposition:
A Practical Guide, 279 – 282.

similar function as laccases in many plants and fungi (Mayer 1987), and is involved in oxidative polymerization of phenolics.

It is virtually impossible to determine individual activities of any of the various phenol-oxidising enzymes in an environmental sample. However, the method by Zimmer & Topp (1998) described here provides an estimate of the overall phenol oxidation capacity. To this end, a suitable phenolic substrate (see Faure et al. 1995) is mixed with the sample, and the change in absorbance resulting from the release of coloured oxidation products is followed over time. Since the oxidation products of phenolic compounds are not clearly defined, no specific extinction coefficient can be determined. The method therefore yields only relative phenol oxidation capacity (ΔA mg^{-1} h^{-1}).

2. EQUIPMENT AND MATERIALS

2.1. Equipment

- Homogenizer (e.g., electronic disperser or mortar and pestle for leaf litter; rotation grinder, or ultrasonic disintegrator for gut and faeces samples)
- Incubation tubes: glass tubes with screw caps (15—20 ml) for leaf litter and plastic reaction tubes (1.5 ml) for gut and faeces samples
- Analytical balance
- Shaker
- Centrifuge (10000 g)
- Micro-pipettes (100—1000 µl; 10—100 µl)
- Plastic cuvettes
- Spectrophotometer (340 nm)

2.2. Materials

- Leaf litter collected in the field
- Dissected guts of detritivores having fed on leaf litter (gut epithelium preferably removed)
- Faeces of detritivores having fed on leaf litter

2.3. Chemicals and Solutions

- 0.05 M KNa phosphate buffer: 415 ml 0.1 M KH_2PO_4 + 85 ml 0.1 M Na_2HPO_4 + 500 ml distilled water, pH 6.2; if prepared accurately, the pH does not need to be adjusted.
- 50 mM catechol in 0.05 M KNa phosphate buffer, pH 6.2.
- Depending on the phenol oxidase investigated, phenolic substrates other than catechol may be more appropriate (e.g., gallic acid, DOPA, tyrosine, syringaldazine). The wavelength for photometric determination of oxidation

products may have to be adjusted accordingly (Faure et al. 1995, Zimmer & Topp 1998).

• Depending on the phenolic substrate, the addition of up to 20% ethanol may be necessary to dissolve the substrate (Faure et al. 1995, Zimmer & Topp 1998).

3. EXPERIMENTAL PROCEDURES

3.1. Extraction of Microbial Enzymes

1. Weigh samples of leaf litter (corresponding to 50—100 mg dry mass), dissected guts (5—10 mg), or faeces (5—10 mg).
2. Determine dry mass:fresh mass ratios to estimate dry mass of samples from fresh mass.
3. The appropriate method of enzyme extraction depends on the source of enzymes; with extracellular enzymes, thoroughly chopping up samples with a homogenizer is sufficient for accurate measurement of enzyme activity; with cell-bound enzymes, additional sonication is recommended to release enzymes into the supernatant.
4. Homogenize litter samples in 10 ml of 0.05 M phosphate buffer, or gut or faeces samples in 1 ml of 0.05 M phosphate buffer. Homogenization must be done on ice to avoid thermal denaturation of enzymes.
5. Homogenates may be stored frozen (–20 °C) until used for assays.
6. Centrifuge suspensions (5 min; ca. 10000 g, depending on the available centrifuge and reaction tubes; 4 °C).

3.2. Determination of Phenol Oxidase Activity

1. Add 100 µl of the supernatant to 900 µl of 50 mM catechol solution and mix thoroughly.
2. Follow change in absorbance (ΔA) at 340 nm at 1-min intervals for the first 10 min.
3. Determine relative catechol oxidation as mean ΔA per min by linear regression analysis.
4. Calculate relative phenol oxidase activity (ΔA mg^{-1} h^{-1}) as:

$$phenol\ oxidase\ activity = \frac{\Delta A \cdot 60 \cdot dilution\ factor}{sample\ dry\ mass} \tag{38.1}$$

where the dilution factor = 100 for litter and 10 for guts and faeces.

4. REFERENCES

Boerjan, W., Ralph, J. & Baucher, M. (2003). Lignin biosynthesis. *Annual Review of Plant Biology*, 54, 519-546.

Breznak, J.A. & Brune, A. (1994). Role of microorganisms in the digestion of lignocellulose by termites. *Annual Review of Entomology*, 39, 453-487.

Claus, H. & Filip, Z. (1990). Effects of clays and other solids on the activity of phenoloxidases produced by some fungi and actinomycetes. *Soil Biology and Biochemistry*, 22, 483-488.

Faure, D., Bouillant, M.-L. & Bally, R. (1995). Comparative study of substrates and inhibitors of *Azospirillum lipoferum* and *Pyricularia oryzae* laccases. *Applied and Environmental Microbiology*, 61, 1144-1146.

Hagerman, A.E. & Butler, L.G. (1991). Tannins and Lignins. In: G.A. Rosenthal & M.R. Berenbaum (eds). *Herbivores: Their Interactions with Secondary Plant Metabolites – I: The Chemical Participants* (pp. 355-388). Academic Press. New York.

Harrison, A.F. (1971). The inhibitory effect of oak leaf litter tannins on the growth of fungi, in relation to litter decomposition. *Soil Biology and Biochemistry*, 3, 167-172.

Kubitzki, K. & Gottlieb, O.R. (1984). Phytochemical aspects of angiosperm origin and evolution. *Acta Botanica Neerlandica*, 33, 457-68.

Ljungdahl, L.G. & Eriksson, K.-E. (1985). Ecology of microbial cellulose degradation. *Advances in Microbial Ecology*, 8, 237-299.

Mayer, A.M. (1987). Polyphenol oxidases in plants – recent progress. *Phytochemistry*, 26, 11-20.

Rabinovich, M.L., Bolobova, A.V. & Vasil'chenko, L.G. (2004) Fungal decomposition of natural aromatic structures and xenobiotics: A review. *Applied Biochemistry and Microbiology*, 40, 1-17.

Sarkanen, K.V. & Ludwig, C.H. (1971). *Lignins – Occurrence, Formation, Structure and Reactions*. Wiley & Sons. New York.

Savoie, J.-M. & Gourbière, F. (1989). Decomposition of cellulose by the species of the fungal succession degrading *Abies alba* needles. *FEMS Microbial Ecology*, 62, 307-314.

Summers, C.B. & Felton, G.W. (1994). Prooxidant effects of phenolic acids on the generalist herbivore *Helicoverpa zea* (Lepidoptera: Noctuidae): potential mode of action for phenolic compounds in plant anti-herbivore chemistry. *Insect Biochemistry and Molecular Biology*, 24, 943-953.

Swain, T. (1979). Tannins and lignins. In: G.A. Rosenthal & D.H. Janzen (eds). *Herbivores: Their Interactions with Secondary Plant Metabolites – I: The Chemical Participants* (pp. 657-718). Academic Press. San Diego.

Thurston, C.F. (1994). The structure and function of fungal laccases. *Microbiology*, 140, 19-26.

Wood, D.A. (1980). Production, purification and properties of extracellular laccase of *Agaricus bisporus*. *Journal of General Microbiology*, 117, 327-338.

Zimmer, M. & Topp, W. (1998). Nutritional biology of terrestrial isopods (Isopoda: Oniscidea): Copper revisited. *Israel Journal of Zoology*, 44, 453-462.

CHAPTER 39

PROTEINASE ACTIVITY: AZOCOLL AND THIN-LAYER ENZYME ASSAY

MANUEL A.S. GRAÇA[1] & FELIX BÄRLOCHER[2]

[1]Departamento de Zoologia, Universidade de Coimbra, 3004-517 Coimbra, Portugal; [2]63B York Street, Dept. Biology, Mt. Allison University, Sackville, N.B., Canada, E4L 1G7.

1. INTRODUCTION

Senescent leaves are rich in structural polysaccharides. Invertebrates can gain access to this energy source by (a) ingesting leaves partially digested by fungi, (b) relying on gut endosymbionts, and/or (c) using fungal enzymes in their own gut (Bärlocher 1982, Graça 1993). While carbon is thus readily available in senescent leaves, nitrogen tends to be scarce, even relative to the much lower demand compared to carbon (Klug & Kotarski 1980, Bernays 1981). In addition, nitrogen accessibility to invertebrates is lowered by the presence of plant polyphenolics that remain active after senescence and complex proteins, where most of the cellular nitrogen is located (Chapters 14 and 15; MacManus et al. 1985, Waterman & Mole 1994). Leaves with low nitrogen and high polyphenolic content are therefore a low-quality food resource for detritivores.

Invertebrate detritivores with low mobility such as tipulid larvae cannot afford to reject low-quality food. Natural selection should have favoured adaptations allowing them to overcome the protein-masking effects of polyphenolics and use N resources more efficiently than highly mobile detritivores (Bärlocher & Porter 1986). This seems to have been achieved by alkaline protein digestion; the pH of the hindgut of some immature tipulid shredders can reach or exceed 10.5 to 11 (Martin et al. 1980, Sinsabaugh et al. 1985, Bärlocher & Porter 1986). Since phenol-protein complexes are less stable under alkaline conditions (Swain 1979, Bärlocher et al. 1989), the pH regime of the midgut allows these tipulids to digest protein in the presence of polyphenolics. However, this is only possible if proteinases remain active at high pH. This has in fact been demonstrated for several tipulid proteinases (Martin et al. 1980, Sinsabaugh et al. 1985, Bärlocher & Porter 1986, Graça & Bärlocher 1998). Given that enzymes active against polyssacharides reach their maximum activity at

M.A.S. Graça, F. Bärlocher & M.O. Gessner (eds.), Methods to Study Litter Decomposition: A Practical Guide, 283 – 288.

acidic to circum-neutral conditions, it is not surprising that polysaccharide and protein digestion in the tipulid gut occur at separate locations characterized by different pH values (Martin et al. 1980, Bärlocher & Porter 1986; Fig. 39.1).

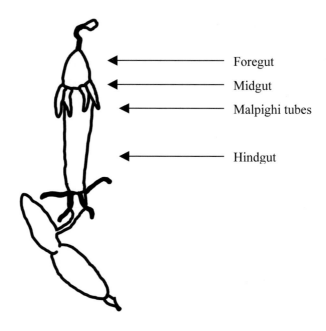

Foregut

Midgut

Malpighi tubes

Hindgut

Fig. 39.1. Schematic representation of a tipulid gut (adapted from Martin et al. 1980)

Here, we describe two simple methods to estimate generalized proteolytic activity. Proteinases can be subdivided on the basis of the peptide linkage they attack (e.g., serine proteinases, cysteine proteinases, etc.). The presented assays measure the combined effects of all these proteinases.

The first assay involves the use of azocoll. Azocoll is a suspension of powdered cow hide to which a bright-red dye is attached (azo dye-bound collagen) (Oakley et al. 1946). The cow hide contains the usual assortment of peptide linkages characteristic of all proteins. When a proteolytic enzyme breaks one of these linkages, the bound dye is released into solution. The rate of this release (determined by measuring absorbance of the filtrate at 520 nm) is used to measure overall proteolytic activity. The described protocol follows Martin et al. (1980).

Wilkström et al. (1981, 1982) described another method to measure the activity of proteinases, which is known as the "thin-layer enzyme assay". It takes advantage of the property of proteins to adsorb to hydrophobic solid surfaces as thin layers, increasing wettability of these surfaces. Polystyrene Petri dishes are hydrophobic surfaces which can be coated with a protein solution. Application of a gut extract containing proteinases will cause protein digestion, and the consequent decrease in wettability can be visualized as a decrease in the condensation of water vapour. The magnitude of protein digestion is proportional to the zone of reduced wettability, expressed as D^2 (squared diameter) of the affected area.

2. EQUIPMENT, CHEMICALS AND SOLUTIONS

2.1. Equipment and Material

- Immature specimens of *Tipula* sp.
- Homogenizer or a manual tissue grinder
- Centrifuge (14000 *g*)
- Sephadex G-25M column
- Spectrophotometer (520 nm)
- Polystyrene Petri dishes
- Microsyringe (10 µl)
- Cork borer
- Fibreglass filter or membrane filter
- Oven (37 °C)
- Pan with hot water (50 °C)

2.2. Chemicals and Solutions

- Ethanol (95%)
- Sterile deionized water
- Bovine Serum Albumin (BSA)
- Agar
- Azocoll
- HCl (0.01 M)
- Buffers: 0.1 M acetate (pH 5.1, 5.6), 0.1 M phosphate (pH 6, 6.5, 7.0, 7.5), 0.1 M tris(hydroxymethyl) aminomethane (Tris; 8.0, 8.5, 9.0), 0.1 M carbonate (pH 9.5, 10.0, 10.5, 11.0), 0.1 M phosphate (11.5, 12.0, 12.5)

3. EXPERIMENTAL PROCEDURES

3.1. Sample Preparation

1. Take live immature specimens of a tipulid shredder. Place in freezer for a few minutes. Open the specimens, and remove the entire gut from the anterior to the posterior end.
2. Homogenize the desired gut section (e.g., midgut, hindgut; combine sections of 3—10 individuals) in 2.5 ml of sterile deionized water.
3. Centrifuge the homogenized gut at 14000 g for 4 min.
4. Desalt the supernatant by passing it through a Sephadex G-25M column and store at 4 °C until needed.
5. Optionally, determine protein content of the extract (see Chapter 9) to express results per mg protein.

3.2. Azocoll

1. Mix 0.5 ml of gut extract with 0.5 ml of buffer and add 25 mg azocoll.
2. Incubate at 37 °C for 15—90 min, depending on enzyme activity.
3. Terminate reaction by adding 3.0 ml of 0.01 M HCl.
4. Pass through fibreglass filter.
5. Measure absorbance at 520 nm.
6. Express results as amount of enzyme required to bring about a change of absorbance at 520 nm under the conditions of the assay (e.g., number of 0.001 absorbance units per minute per gut section at pH = 8 and incubation at 37 °C).

3.3. Thin-Layer Enzyme Assay

1. Clean the internal surface of polystyrene Petri dishes with 95% ethanol and dry in an oven at 37 ºC.
2. Coat the internal surface of the Petri dishes with protein (3 ml of BSA, 1 mg ml^{-1} for 30 min). Discard the excess protein solution and gently rinse with sterile deionized water. Dry at room temperature or in an oven at 37 ºC.
3. Prepare an agar solution (2%) and apply 10 ml over the coated Petri dishes. Allow to gel at room temperature.
4. Cut 3—4 mm diameter wells in the agar and apply 10 µl of desalted gut extract.
5. Incubate the Petri dishes for 18 h at 37 °C.
6. Remove the agar manually, gently wash with distilled water and dry.

7. Expose the Petri dish to water vapour from a hot (50 °C) water pan. Digestion of protein is visible as condensation of water vapour due to reduced surface wettability (Fig. 39.2).

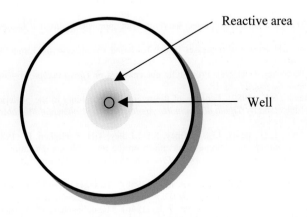

Figure 39.2. Decreased wettability of the internal surface of a polystyrene Petri dish, caused by digestion of proteins.

8. Determine the size of the affected area or an equivalent measure (e.g. squared diameter).
9. To estimate proteolytic activity as a function of pH, buffers can be used to adjust the pH of the agar to values similar to the ones in the gut or environment.

4. REFERENCES

Bärlocher, F. (1982). The contribution of fungal enzymes to the digestion of leaves by *Gammarus fossarum* Koch (Amphipoda). *Oecologia*, 52, 1-4.

Bärlocher, F. & Porter, C.W. (1986). Digestive enzymes and feeding strategies of three stream invertebrates. *Journal of the North American Benthological Society*, 5, 58-66.

Bärlocher, F., Tibbo, P.G. & Christie, S.H. (1989). Formation of phenol-protein complexes and their use by two stream invertebrates. *Hydrobiologia*, 173, 243-249.

Bernays, E.A. (1981). Plant tannins and insect hervivores: an appraisal. *Ecological Entomology*, 6, 353-360.

Graça, M.A.S. (1993). Patterns and processes in detritus-based stream systems. Limnologica, 23, 107-114.

Graça, M.A.S. & Bärlocher, F. (1998). Proteolytic gut enzymes in *Tipula caloptera* – interaction with phenolics. *Aquatic Insects*, 21, 11-18.

Klug, M.J. & Kotarski, S.F. (1980). Ecology of the microbial microbiota in the posterior hind gut of larval stages of the crane fly *Tipula abdominalis*. *Applied and Environmental Microbiology*, 40, 408-416.

MacManus, J.P., Davis, K.G., Beart, J.E., Goffney, S.H., Lilley, T.H. & Haslam, E. (1985) Polyphenol interaction. Part 1. Introduction: Some observations on the reversible complexation of polyphenols with proteins and polysaccharides. *Journal of the Chemical Society, Perkin Transactions*, II, 1429-1438.

Martin, M.M., Martin, J.S., Kukor, J.J. & Merrit, R.W. (1980). The digestion of protein and carbohydrate by the stream detritivore *Tipula abdominalis* (Diptera, Tipulidae). *Oecologia*, 46, 360-364.

Oakley, C.L., Warrack, G.H. & van Heyningen, W.E. (1946). The collagenase (K Toxin) of *Cl. welchii* Type A. *Journal of Pathology*, 58, 229-235.

Sinsabaugh, R.L., Linkins, A.E. & Benfield, E.F. (1985). Cellulose digestion and assimilation by three leaf-shredding aquactic insects. *Ecology*, 66, 1464-1471.

Swain, T. (1979). Tannins and lignins. In: G. A. Rosenthal, & D. H. Janzen (eds.), *Herbivores: Their Interactions with Secondary Plant Metabolites* (pp. 657-682). Academic Press. New York.

Waterman, P.G. & Mole, S. (1994). *Analysis of Phenolic Plant Metabolites*. Blackwell Scientific Publications. London.

Wikström, M., Elwing, H. & Linde, A. (1981). Determination of proteolytic activity: a sensitive and simple assay utilizing substrate adsorbed to a plastic surface and radial diffusion in gel. *Analytical Biochemistry*, 118, 240-246.

Wikström, M.B., Elwing, H. & Möller, A.J.R. (1982). Proteins adsorbed to a hydrophobic surface used for determination of proteolytic activity. *Enzyme and Microbial Technology*, 4, 265-268.

PART 5

DETRITIVOROUS CONSUMERS

CHAPTER 40

MAINTENANCE OF SHREDDERS IN THE LABORATORY

FERNANDO COBO

Departmento de Biología Animal, Facultad de Biología, Universidad de Santiago de Compostela, 15782, Santiago de Compostela, Spain.

1. INTRODUCTION

Shredders are aquatic invertebrates whose mouth parts are adapted for feeding on large particles of organic matter such as decomposing leaves. Most shredders from temperate areas are insects (primarily Plecoptera, Tipulidae, Limnephilidae and other Trichoptera) and crustaceans (Amphipoda, Isopoda). Shredders can be grown and maintained in the laboratory if they are provided with adequate food and if the water is kept cool, pure, and well oxygenated. The latter two conditions are easy to achieve in an aquarium with a good filter and adequate aeration. Low temperature is more difficult to attain if expensive equipment such as temperature-controlled rooms or chambers, large cooled incubators or low-temperature aquaria are unavailable. Some species also need water flow to ensure high survivorship, which poses an additional constraint in the design of a maintenance system.

Studying invertebrates in the laboratory provides biologists with an opportunity to work in a controlled environment. A variety of techniques have been described (Allegret & Denis 1972, Armitage & Davis 1989, MacKay 1981, Craig 1993). They range from simple Petri dishes to recirculating streams with complex pumps and cooling systems. Several systems have been constructed mainly for taxonomic or behavioural studies (e.g. Wiggins 1959, Resh 1972, Wiley & Kholer 1980, Smith 1984, Keiper & Foote 1996). If reproduction is successful, large numbers of individuals can be reared and then used in experiments aimed at assessing the involvement of shredders in decomposition (e.g., determination of consumption rates, food preferences, gut enzyme activities). Moreover, some stream invertebrates can only be accurately identified as adults or from exuviae. Consequently, larval stages often have to be reared if identification to the species level is sought (Philipson 1953, Hiley 1969, Bjarnov & Thorup 1970).

M.A.S. Graça, F. Bärlocher & M.O. Gessner (eds.), Methods to Study Litter Decomposition: A Practical Guide, 291–296.

This chapter describes an inexpensive versatile laboratory system for maintaining and rearing stream macroinvertebrates. The system has been used successfully to grow Trichoptera, Plecoptera, Ephemeroptera and Chironomidae from eggs to adults. The reared specimens of these taxa have been collected in cold, fast flowing waters.

2. DESIGN AND MATERIALS

A refrigerator for low-temperature storage of meals cooked in advance as commonly used in catering is used to keep the invertebrates in a temperature-controlled environment (Fig. 40.1). The refrigerator should have a stainless-steel tray that can be used as an aquarium. The tray should be located on the top of the refrigeration circuit that is controlled by a thermostat (Fig. 40.1 A) and preferably be covered by a plexiglass case.

Two movable front panels allow manipulation of the invertebrates. The plexiglass enclosure creates air space for emerged adults. At the top of the plexiglass enclosure, two fluorescent tubes of the "day-light" type (8 W) are installed (Fig. 40.1 B). The electrical circuits for the thermostat and fluorescent tubes should be kept separate so that the tubes can be connected to a programming device controlling the light-dark cycle.

Water leaves the stainless-steel tray by a plastic tube passing through a hole drilled in the tray and the case. A second tube enters the refrigerator through another hole drilled in the plexiglass cover. The tube is blocked at its end, perforated along the section allocated inside the refrigerator and fixed above the water level so that incoming water cascades over the tray, creating turbulence and thus ensuring oxygenation (Fig. 40.2). An aquarium pump provided with a filtering system circulates and purifies the water (Fig. 40.1 C).

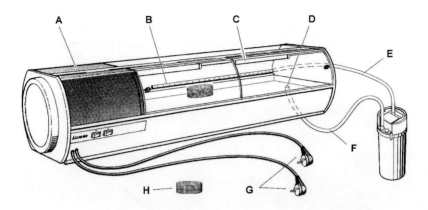

Figure 40.1. Schematic representation of the closed culture system. A: Refrigerator; B: Tube sealed at the end and with small holes inside the chamber; C: Fluorescent tubes. D: Hole in the tray for water exit; E and F: outflow and inflow tubes for water circulation and water pump with filtering system; G: Power connections (light and refrigerator); H: Individual enclosure.

Additionally, small enclosures (Fig. 40.1 H) can be used to rear invertebrates individually within the refrigerator. Enclosures can be constructed using a strong coarse plastic mesh as a framework and a fine plastic mesh lining the inner sides. The size of the inner mesh can range from 250—500 μm, depending on size of the individuals. The top of the enclosures can be covered with plastic lids or Petri dishes, in case individuals are to be collected individually from each chamber. The enclosures are placed directly over the steel tray.

It is often important to keep individual invertebrates separate to avoid predation; this allows maintaining several species simultaneously. Moreover, some pupae and final instar larvae can use the plastic mesh of enclosures to climb out of the water and emerge. The plastic lids or Petri dishes prevent adults from leaving the enclosures. When rearing Trichoptera, materials suitable for the construction of the larval case have to be added (e.g. sand, small pieces of wood, moss).

The water used to rear invertebrates should ideally be filtered stream water. However, when this is not possible, artificial stream water can be used. Many formulas have been proposed (e.g. Graça et al. 1993). The following composition has ensured high survival rates for many species in our laboratory: 35 mg NaCl, 2 mg KH_2PO_4, 61.5 mg $MgSO_4$, 36 mg $CaCl_2$, 5 mg $NaHCO_3$ and 1.6 mg $FeCl_3$ per litre.

Routine maintenance of the system consists of adding distilled water (UV irradiated or autoclaved; pH adjusted to values observed in streams). The stainless-steel tray and the filter must be cleaned, and the water must be renewed about every two months.

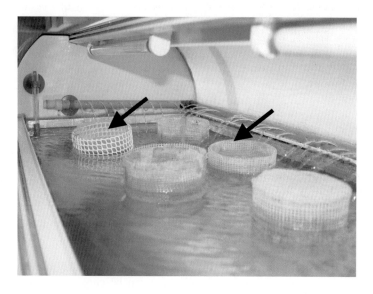

Figure 40.2. Enclosures (arrows) with invertebrates in the tray of the culture system

3. REFERENCES

Allegret, P. & Denis, C. (1972). Dispositif pour l'élevage d'insectes aquatiques a température constante. *Annales d'Hydrobiologie*, 3, 65-67.

Armitage, P. & Davies, A. (1989). A versatile laboratory stream with examples of its use in the investigation of invertebrate behaviour. *Hydrobiological Bulletin*, 23, 151-160.

Bjarnov, N. & Thorup, J. (1970). A simple method for rearing running-water insects, with some preliminary results. *Archiv für Hydrobiologie*, 67, 201-209.

Craig, D.A. (1993). Hydrodynamic considerations in artificial stream research. In G.A. Lamberti, & A.D. Steinman (eds.), *Research in Artificial Streams: Applications, Uses and Abuses* (pp. 324-327). *Journal of the North American Benthological Society*, 12, 313-384.

Graça, M.A.S., Maltby, L. & Calow, P. (1993). Importance of fungi in the diet of *Gammarus pulex* (L.) and *Asellus aquaticus* (L.): II. Effects on growth, reproduction and physiology. *Oecologia*, 96, 304-309.

Hiley, P.D. (1969). A method of rearing Trichoptera larvae for taxonomic purposes. *Entomologist's Monthly Magazine*, 105, 278-279.

Keiper, J.B. & Foote, B. A. (1996). A simple rearing chamber for lotic insect larvae. *Hydrobiologia*, 339, 137-139.

MacKay, R.J. (1981). A miniature laboratory stream powered by air bubbles. *Hydrobiologia*, 83, 383-385.

Philipson, G.N. (1953). A method of rearing trichopterous larvae collected from swift-flowing waters. *Proceedings of the Royal Entomological Society of London, Series A*, 28, 15-16.

Resh, V.H. (1972). A technique for rearing caddisflies (Trichoptera). *Canadian Entomologist*, 104, 1959-1961.

Smith, M.H. (1984). Laboratory rearing of stream-dwelling insects. *Antenna*, 8, 67-69.

Wiggins, G.B. (1959). A method of rearing caddisflies (Trichoptera). *Canadian Entomologist*, 91, 402-405.

Wiley, M.J., & Kholer, S.L. (1980). Positioning changes of mayfly nymphs due to behavioural regulation of oxygen consumption. *Canadian Journal of Zoology*, 58, 618-622.

CHAPTER 41

FEEDING PREFERENCES

CRISTINA CANHOTO[1], MANUEL A.S. GRAÇA[1] & FELIX
BÄRLOCHER[2]

[1]*Departamento de Zoologia, Universidade de Coimbra, 3004-517 Coimbra, Portugal;* [2]*63B
York Street, Dept. Biology, Mt. Allison University, Sackville, N.B., Canada E4L 1G7.*

1. INTRODUCTION

When given the choice, consumers often preferentially feed on some food items
while ignoring others. This can profoundly influence species diversity at lower
trophic levels, and may give important clues to co-evolution between predator and
prey. Not surprisingly, there is a vast body of literature on food choices in
freshwater, marine and terrestrial consumers (e.g. Steinberg 1985, Graça et al. 1993,
Bonkowsky et al. 2000). The general design of most of these studies is
straightforward: in a container (representing one replicate, or a block), a consumer is
given a choice among several food types. The amount of the various items
consumed is then estimated, generally by comparing area or mass before and after
exposure to consumers. The difference is assumed to be due to feeding. When mass
loss can occur in the absence of feeders, it must be estimated with separate control
replicates. These are treated identically, except that they do not contain consumers.

Statistical analysis of food choice experiments presents several problems
(Bärlocher 1999). To begin with, mass losses in controls have been used to calculate
'correction factors', i.e. average losses occurring in the absence of consumers were
subtracted from changes found in the presence of the consumers. The resulting
values were assumed to represent consumption and were used in statistical analyses.
This approach ignores variability between control replicates and should hence be
abandoned: when two independent variables are added or subtracted, the total
variance corresponds to the sum of the variances of the two variables. The
consequences will be negligible when the variance among control replicates is much
smaller than the variance among treatment replicates. If the variance in controls is
sizeable, one can maintain overall variance by forming random pairs between
control and experimental data, and determine the differences. To avoid random
pairing, a test can be based on the comparison of the means of two multivariate
vectors represented by controls and treatments (Manly 1986, 1993, Krebs 1999).

*M.A.S. Graça, F. Bärlocher & M.O. Gessner (eds.), Methods to Study Litter Decomposition:
A Practical Guide, 297 – 302.*
© 2005 *Springer. Printed in The Netherlands.*

The second problem of food choice experiments concerns 'independence of variables'. When two or more food items are presented in the same container, consumption of one cannot be assumed to be independent of consumption of the other(s). Conventional statistical tests are therefore inappropriate. As alternative, one can use Friedman's test. It is a modification of Fisher's permutation test: actual measurements are first converted to ranks (to maintain interdependencies, food items within each feeding container are ranked separately). This conversion to ranks was necessary at the time when the test was developed because of a lack of powerful computers. It results in a loss of power of the test, making it less likely that significant differences can be identified.

An alternative approach to the multiple choice experiments is to run tests with only two food items presented in each experimental replicate, and prepare replicates with all pair-wise combinations of tested foods. This approach was first suggested by Petersen & Renaud (1989) and used by Friberg & Jacobsen (1994) and Graça et al. (2001). This alternative allows the ranking of food items in terms of preference. However, the problem with this approach is the high number of trials needed; for instance, if one wants to evaluate preference of an invertebrate for 7 food types, 21 pair trials must be run. In addition, the sequence of preferences based on pair-wise choices may not correspond to the sequence when all foods are offered simultaneously (Manly 1993).

Detritivores are selective feeders and their preferences have been related with leaf senescence (Yeates & Barmuta 1999), nutrient content of food and microbial conditioning (Graça et al. 2001) and presence of unpalatable or indigestible compounds (Target et al. 1986, Canhoto & Graça 1999). In this chapter we describe an approach to assess feeding preferences of shredders from streams; it can easily be adapted for terrestrial, marine or other freshwater consumers.

2. EQUIPMENT AND MATERIALS

- Oven (40—50 °C)
- Analytical balance ($\pm$ 0.01 mg)
- Vacuum pump and filtering system
- Plastic cups (approx. 10 cm high, 8 cm diameter; total number = 48)
- Aeration system (plastic tube, pipette tips, compressor)
- Cork borer
- Large litter bags (10 × 14 cm; 0.5 mm mesh size)
- Small litter bags (2 × 2 cm; 0.5 mm mesh size)
- Clips
- Pins with coloured heads
- Culture racks for biological samples (e.g. Nunc®, 5 × 4 chambers rack)
- Aluminium foil
- Stream water
- Stream sand devoid of organic matter (place at 500 °C for 8 h)
- Fibre glass, membrane or paper filters

- Food items of 2 or more categories. These could be different leaf species, or conditioned vs. unconditioned leaves.
- Shredders (e.g. *Sericostoma* sp., *Tipula* sp., *Gammarus* sp.)

3. EXPERIMENTAL PROCEDURES

3.1. General

1. Place ~4 g of leaves in litter bags and incubate in a stream for an appropriate time (e.g. 3 weeks) to "condition" the leaves, i.e. to allow microbial colonization and changes in physical and chemical tissue quality.
2. After incubation, remove the leaves from the litter bags and rinse under running tap water.
3. Cut pairs of discs (one from each side of the main vein) with a cork borer ($\varnothing$ = 1 cm). One disc of each pair will be used as control and the other will be offered to the invertebrates (Fig. 41.1). Prepare material for ca. 20 replicates.

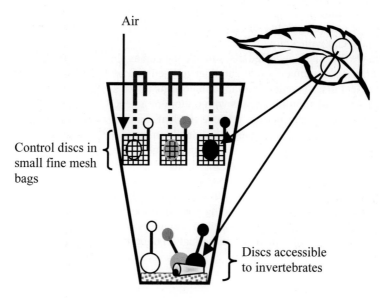

Air

Control discs in small fine mesh bags

Discs accessible to invertebrates

Figure 41.1. Schematic representation of a feeding cup during a food choice experiment with 3 pairs of discs from different leaf types.

4. Filter the stream water and add ~200 ml to each cup.
5. Cover the bottom of containers with sand (~5 mm).
6. Prepare the aeration system as shown in Fig. 41.2.

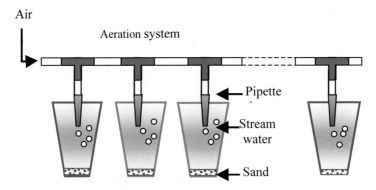

Figure 41.2. Series of chambers for feeding experiments.

3.2. Selection Among Several Food Items

1. Mark the leaf discs of each leaf type with coloured pins (e.g. sp. 1 – white head; sp. 2 – blue head; sp. 3 – green head).
2. Place one control disc of each pair inside a small litter bag that will be attached to the cup with a clip.
3. Place the other identified discs inside the cup as shown in Fig. 41.1.
4. Allocate one invertebrate shredder to each cup and allow feeding until one of the leaf discs is reduced to about half of its initial size.
5. Make small aluminium pans with bottom of a pencil, label and weigh them.
6. After the feeding period, retrieve the leaf material (discs exposed to shredders and control discs) and place individually in the aluminium pans.
7. Dry the discs in the oven for 48 h and weigh to the nearest 0.01 mg.
8. Proceed in the same way with the shredders.
9. Estimate individual consumption for each leaf type (mg) as the difference between the weight of the control leaf disc (L_c) (whose initial weight is assumed to be similar to its corresponding disc from the same leaf) and the weight of the corresponding leaf disc exposed to shredders (L_e). Results can be expressed as mg dry mass consumed per individual (I) over the feeding time (e.g. in days):

$$C = \frac{L_c - L_e}{I \cdot t} \qquad (41.1)$$

3.3. Statistical Analysis

1. Evaluate food preferences by Friedman's test, which is based on ranks.
2. Alternatively, evaluate actual consumption values by a permutation test, using Resampling Stats or a similar program (see Chapter 43).
   a. Define a test statistic S that measures differences in consumption rates of the various food items, e.g., the sum of squared deviations from average consumption. Calculate the value of S for the original data.

b. Within each container, randomly assign measured consumption values to the available food items. For each of these permutations, determine the new value of S. Repeat this 10000 times; this gives the distribution of all possible values of S.

c. How extreme is the original value of S compared to the entire distribution? Determine the proportion of S values that is at least as large as the original value of S; this proportion corresponds to the p-value of traditional statistical methods.

4. FINAL REMARKS

The paired design described above avoids underestimates of the variability of consumption rates. If this design is not possible, individual leaf discs can be dried and weighed. Before the experiment the discs should be rehydrated and exposed to shredders. A series of leaf discs without shredders should be used as controls. In this case the initial weigh of each disc is known. Consumption can be estimated as:

$$C = \frac{W_i - W_f \cdot F}{I \cdot t} \tag{41.2}$$

Where: C = consumption; F = correction factor given by the ratio of initial to final mass of a set of control discs; I = invertebrate dry mass; t = duration of feeding trial; W_i = initial leaf disc weight; W_f = final leaf disc weight. The use of an average correction factor will underestimate the variance of C. If the ratio of initial to final weight in control discs is highly variable, a comparison of two multinomial vectors (weight losses in control and in feeding containers) may be more appropriate (Manly 1993).

The approach described here can also be used as to assess feeding rates when consumers are exposed to a single leaf type. Feeding rates of 0.47—1.88 mg of leaves invertebrate^{-1} day^{-1} have been reported for sericostomatid caddis-flies (Feio & Graça 2000, Wagner 1990) feeding on conditioned alder (*Alnus glutinosa*) leaves. Typical feeding rates for stream shredders range from 0.04 to 0.5 mg mg^{-1} day^{-1} (Arsuffi & Suberkropp 1989). Growth rates can also be estimated if sizes of invertebrates are known at the beginning and end of the experiment. In this case periodical (weekly) changes in food and water is indicated.

For terrestrial consumers such as isopods, feeding experiments can be carried out in 5.5 mm diameter Petri dishes. High humidity can be achieved by applying wet filter paper discs to the lid of the Petri dish.

5. REFERENCES

Arsuffi, T.L. & Suberkropp, K. (1989). Selective feeding by shredders on leaf-colonizing stream fungi: comparison of macroinvertebrate taxa. *Oecologia*, 79, 30-37.

Bärlocher, F. (1999). Statistical analysis of feeding experiments: a resampling approach. http://homepage.mac.com/fbaerlocher/noger/Feeding_preferencesPM

Bonkowsky, M., Griffiths, B.S. & Ritz, K. (2000). Food preferences of earthworms for soil fungi. *Pedobiologia*, 44, 666-676.

Canhoto, C.M. & Graça, M.A.S. (1999). Leaf barriers to fungal colonization and shredders (*Tipula lateralis*) consumption of decomposing *Eucalyptus globulus*. *Microbial Ecology*, 37, 163-172.

Feio, M.J. & Graça, M.A.S. (2000). Food consumption by the larvae of *Sericostoma vittatum* (Trichoptera), an endemic species from the Iberian Peninsula. *Hydrobiologia*, 439, 7-11.

Friberg, N. & Jacobsen, D. (1994). Feeding plasticity of two detritivore-shredders. *Freshwater Biology*, 32, 133-142.

Graça, M.A.S., Maltby, L. & Calow, P. (1993). Importance of fungi in the diet of *Gammarus pulex* and *Asellus aquaticus*. I: feeding strategies. *Oecologia*, 93, 139-144.

Graça, M.A.S., Cressa, C., Gessner, M.O., Feio, M.J., Callies, K.A. & Barrios, C. (2001). Food quality, feeding preferences, survival and growth of shredders from temperate and tropical streams. *Freshwater Biology*, 46, 947-957.

Krebs, C.J. (1999). *Ecological Methodology*. Benjamin/Cummings. Menlo Park, CA.

Manly, B.F.J. (1986). *Multivariate Statistical Methods*. Chapman & Hall. London.

Manly, B.F.J. (1993). Comments on design and analysis of multiple-choice feeding-preference experiments. *Oecologia*, 93, 149-152.

Petersen, C.H. & Renaud, P.E. (1989). Analysis of feeding preference experiments. *Oecologia*, 80, 82-86.

Steinberg, P.D. (1985). Feeding preferences of *Tegula funebralis* and chemical defenses of marine brown algae. *Ecological Monographs*, 55, 333-349.

Target, N.M., Target, T.E., Vrolijk, N.H. & Ogden, J.C. (1986). The effect of macrophyte secondary metabolites on feeding preferences of the herbivorous parrotfish *Sparisma radians*. *Marine Biology*, 92, 141-148.

Wagner, R. (1990). A laboratory study on the cycle of *Sericostoma personatum* (Kirby & Spence) and light-dark dependency. *Hydrobiologia*, 208, 201-211.

Yeates, L.V. & Barmuta, L.A. (1999). The effects of willow and eucalypt leaves on feeding preference and growth of some Australian aquatic macroinvertebrates. *Australian Journal of Ecology*, 24, 593-598.

PART 6

DATA ANALYSIS

CHAPTER 42

BIODIVERSITY

FELIX BÄRLOCHER

63B York Street, Dept. Biology, Mt. Allison University, Sackville, N.B., Canada E4L 1G7.

1. INTRODUCTION

There is great concern about the ongoing permanent loss of species. One important question is: How will this affect important aspects of ecosystem functioning? Ehrlich & Ehrlich (1981) wrote: "Ecosystems, like well-made airplanes, tend to have redundant subsystems and other 'design' features that permit them to continue functioning after absorbing a certain amount of abuse. A dozen rivets, or a dozen species, might never be missed. On the other hand, a thirteenth rivet popped from a wing flap, or the extinction of a key species involved in the cycling of nitrogen could lead to a serious accident." In recent years, a great number of studies have explored potential relationships between diversity and ecological functions, and tried to fit them into one of several models (for reviews, see Kinzig et al. 2001, Loreau et al. 2002, Wardle 2002). Most investigations dealt with plant species and primary production; a smaller number have investigated microorganisms and decomposition. Some recent studies looked at stream invertebrates (Jonsson & Malmquist 2000), and aquatic hyphomycetes (Bärlocher & Corkum 2003) in relation to leaf decomposition (Covich et al. 2004).

In laboratory studies, the number of species can be controlled. This is generally not the case in field studies, where the number of species is unknown and has to be estimated. These estimates depend crucially on sample size

This section provides an introduction to some concepts that are important when measuring and comparing diversities. An excellent overview is given by Krebs (1999), who has also produced a computer program that automates many of the calculations mentioned in the text. It can be applied, for example, to calculate the diversity of aquatic hyphomycete spores (Chapter 24) or of invertebrates colonizing leaves in decomposition experiments (Chapter 6).

M.A.S. Graça, F. Bärlocher & M.O. Gessner (eds.), Methods to Study Litter Decomposition:
A Practical Guide, 305 – 312.
© 2005 *Springer. Printed in The Netherlands.*

2. ESTIMATING SPECIES-RICHNESS

2.1. Rarefaction

The larger a sample, the greater will be the expected number of species and the lower the evenness (Rosenzweig 1999). If we observe 88 species in a collection of 1500 individuals (community A) and 55 species in a collection of 855 individuals (community B), we do not readily know which community has more species. For a meaningful comparison we have to standardize the sample size. We do this by a method called rarefaction, which was introduced by Sanders (1968). It answers the following question: If a sample had consisted of n instead of N individuals ($n<N$), how many species (s) would have been found? The largest sample in a collection is assumed to have S species distributed among N individuals; all rarefied samples have $n < N$ individuals and $s < S$ species. We can use the following formula:

$$E(\hat{S}_n) = \sum_{i=1}^{S} \left[1 - \frac{\binom{N-N_i}{n}}{\binom{N}{n}} \right]$$ (42.1)

where
$E(\hat{S}_n)$ = expected number of species in a random sample of n individuals
S = total number of species in entire collection
N_i = number of individuals belonging to species i
N = total number of individuals in collection
n = sample size chosen for standardization.

Alternatively, we can determine the expected number of species empirically by repeatedly taking subsamples of the size chosen for standardization. We can simulate this process by using a computer program, such as Resampling Stats (Chapter 43). For example, if a sporulation experiment (Chapter 24) results in a filter with 1064 aquatic hyphomycete conidia and 8 species (Table 2), how many species would be expected in a sample of 250 individuals?

Table 42.1. Fictitious result of a sporulation experiment.

Species	Number of conidia
Anguillospora filiformis	550
Articulospora tetracladia	25
Clavariopsis aquatica	123
Flagellospora curvula	17
Heliscus lugdunensis	222
Lemonniera aquatica	120
Tetracladium marchalianum	5
Tumularia aquatica	2
Total of 8 species	1064

The formula gives a value of 7.1. The same value can be estimated with the simple Resampling Stats program listed below. It defines 8 species by assigning them numbers 1 to 8. The number of individuals belonging to each species is defined by urns: URN 550#1 25#2... implies 550 individuals of species 1, 25 of species 2, etc. The numbers are shuffled, and a sample of 250 is taken. All duplicates (i.e., identical numbers = individuals belonging to the same species) are removed, and the remaining numbers, corresponding to different species, are counted. This is repeated 10000 times. The average gives the expected number of different species when a sample of 250 is taken. The commands to simulate rarefaction with Resampling Stats are as follows:

```
MAXSIZE default 150000
URN 550#1 25#2 123#3 17#4 222#5 120#6 5#7 2#8
REPEAT 10000
SHUFFLE A B
TAKE B 1,250
DEDUP C D
COUNT D>0 E
SCORE E F
END
MEAN F aver
PRINT aver
```

2.2. Species-Area Curve Estimates

Another way to estimate species-richness is to extrapolate the species-area curve for the community. Since the number of species tends to rise with the area sampled, one can fit a regression line and use it to predict the number of species of any size. The species-area relationship generally has the following form:

$$S = c \cdot A^z \qquad (42.2)$$

where
 S = number of species
 c, z = constants
 A = area

Provided we have several samples with known area and species number, we can estimate c and z by non-linear curve-fitting with statistical (e.g. SYSTAT, SAS) or mathematical (e.g. MatLab, Mathematica) software. The samples could then be grouped based on a factor of interest (e.g., fungal conidia in streams bordered by different forest cover), to test whether the values of one group are consistently above or below the estimated species-area curve. Species-area curves have been applied to aquatic hyphomycetes by Gönczöl et al. (2001). This method should not be used for sparsely sampled sites.

2.3. Assuming an Upper Limit

Each habitat supports a limited number of species. We can estimate this upper limit by plotting the number of different species as function of the number of examined individuals or number of samples. The resulting curve often resembles a rectangular hyperbola or saturation-binding curve (also known as Monod or Michaelis-Menten type curve). Fig. 42.1 shows the number of aquatic hyphomycete species found in a stream as a function of the number of monthly samples. In this particular example, the data closely resemble a hyperbola until Month 52, when the number of new species started to rise again. The estimated maximum number of 76 is therefore clearly too low in this case. Alternative methods to extrapolate to true richness in a habitat from a limited number of samples are discussed in Krebs (1999). A recent study compared the performance of six techniques when estimating diversity of sandy beach macroinvertebrates (Foggo et al. 2003); their application to aquatic hyphomycetes is discussed in Bärlocher (2004).

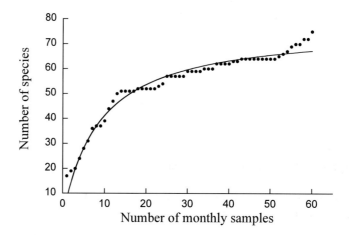

Figure 42.1. Number of identified species with increasing number of samples. Data from Bärlocher (2000), with non-linear curve-fit to rectangular hyperbola.

3. COMPONENTS OF SPECIES DIVERSITY: RICHNESS, HETEROGENEITY, EVENNESS

3.1. Species Richness

In a community with ten equally common species, two randomly collected individuals are unlikely to belong to the same species. In another community with ten species, where 99% of all individuals belong to the same species, two random samples will likely recover the same species. Both communities have the same species richness, which is generally taken to be synonymous with number of species, but the first community is more heterogeneous.

3.2. Heterogeneity

Heterogeneity of a population contains two separate aspects: species richness and evenness. Simpson's index (1949) was the first attempt to combine the two in a single number; it is also known as the 'Repeat Rate', since the index expresses the probability that two organisms selected at random from a population will 'repeat' their classification, that is that they belong to the same species. The repeat rate measure was first used in a German text on cryptography (the science of analyzing and deciphering codes, ciphers and cryptograms) in 1879 (Krebs 1999). For an infinite population, the repeat probability is given by

$$D = \sum p_i^2 \tag{42.3}$$

where
D = Simpson's index
p_i = proportion of species i in community

To convert this probability into a measure of diversity, usually the complement of Simpson's Index $(1 - D)$ is taken:

$$\text{Simpson's index of diversity} = \left\{ \begin{array}{c} \text{probability of picking} \\ \text{two organisms of} \\ \text{different species} \end{array} \right\}$$

Thus,

$$1 - D = 1 - \sum p_i^2 \tag{42.4}$$

Strictly speaking, this formula can only be used for infinite populations (Pielou 1969). For a finite population, the correct estimator is:

$$1 - \hat{D} = 1 - \sum_{i-1}^{s} \left[\frac{n_i(n_i - 1)}{N(N-1)} \right] \qquad (42.5)$$

where
    n_i = number of individuals of species i in sample
    N = total number of individuals in sample
    S = number of species in sample
Applying the formula to the data in Table 42.1 gives a Simpson diversity $(1 - D)$ of 0.662.

The most popular measures of species diversity are based on information theory. The objective is to measure the amount of order (or disorder) present in a system. The underlying question is: How difficult would it be to predict correctly the species of the next individual collected? (In informatics, engineers are interested in correctly predicting the next letter in a message.) This uncertainty can be measured by the Shannon-Wiener function:

$$H' = \sum_{i=1}^{n} p_i \cdot \log_2 p_i \qquad (42.6)$$

where
    H' = index of species diversity = information content of sample (bits/individual)
    S = number of species
    p_i = proportion of total sample belong to the ith species

Information content measures uncertainty: The greater H', the greater the uncertainty. A message such as BBBBB (or a community with only one species) has no uncertainty, and $H' = 0$. The Shannon-Wiener index should only be used on random samples from a large community in which the total number of species is known (Pielou 1969). If this is not the case, the Brillouin index is more appropriate (Krebs 1999). In practice the two indices give nearly identical results, provided the sample is large. For the data in Table 42.1, the Shannon-Wiener index is 1.955 and the Brillouin index 1.930.

3.3. Evenness Measures

The literature on how to measure evenness (or equitability) is vast. Generally, one of the heterogeneity indices is scaled relative to the maximal value it reaches when each species is equally common. Two formulations are common:

$$V = \frac{D}{D_{max}} \tag{42.7}$$

and

$$V' = \frac{D - D_{min}}{D_{max} - D_{min}} \tag{42.8}$$

where
  V, V' = evenness
  D = observed index of diversity
  D_{max} = maximum possible value of index, given S species and N individuals
  D_{min} = minimum possible value of index, given S species and N individuals
The first expression is more commonly used, but the two converge for large samples.
  A wide range of evenness indices has been proposed. Smith & Wilson (1996) prefer the following four: Simpson's, Camargo's, Smith and Wilson's, Modified Nee's index. They all assume that the total number of species is known, which is almost never true. The evenness ratio is therefore always an overestimate. Only Simpson's index is briefly introduced here. Simpson's measure of heterogeneity reaches a maximum when all species are equally abundant ($p = 1/s$); therefore:

$$\hat{D}_{max} = \frac{1}{S} \tag{42.9}$$

  It follows that the maximum possible value of the reciprocal of Simpson's index is always equal to the number of species observed in the sample. Simpson's index of evenness is therefore defined as:

$$E_{1/D} = \frac{1/\hat{D}}{S} \tag{42.10}$$

where
  $E_{1/D}$ = Simpson's measure of evenness
  $\hat{D}$ = Simpson's index
  S = number of species in sample
This index is relatively little affected by rare species. For the data in Table 42.1, $E_{1/D}$ is 0.370.

REFERENCES

Bärlocher, F. (2000). Water-borne conidia of aquatic hyphomycetes: seasonal and yearly patterns in Catamaran Brook, New Brunswick, Canada. *Canadian Journal of Botany*, 78, 157-167.

Bärlocher, F. (2004). Freshwater fungal communities. In: J. Dighton, P. Oudemans, & J. White (eds.), *The Fungal Community*. 3rd ed. Dekker. New York.

Bärlocher, F. & Corkum, M. (2003). Nutrient enrichment overwhelms diversity effects in leaf decomposition by stream fungi. *Oikos*, 101, 247-252.

Covich A.P., Austen, M.C., Bärlocher, F., Chauvet, E., Cardinale, B.J., Biles, C.L., Inchausti, P., Gessner, M.O., Dangles, O., Statzner, B., Solan, M., Moss, B.R. & Asmus, H. (2004). The role of biodiversity in the functioning of freshwater and marine benthic ecosystems: Current evidence and future research needs. *BioScience*, 54, 767-775.

Ehrlich, P.R. & Ehrlich, A.H. (1981). 1981. *Extinction: The Causes and Consequences of the Disappearance of Species*. Random House. New York.

Foggo, A., Attrill, M.J., Frost, M.T. & Rowden, A.A. (2003). Estimating marine species richness: an evaluation of six extrapolative techniques. *Marine Ecology Progress Series*, 248, 15-26.

Gönczöl, J., Révay, A. & Csontas, P. (2001). Effect of sample size on the detection of species and conidial numbers of aquatic hyphomycetes collected by membrane filtration. *Archiv für Hydrobiologie*, 150, 677-691.

Jonnson, M. & Malmqvist, B. (2000). Ecosystem process rates increases with animal species richness: evidence from leaf-eating, aquatic insects. *Oikos*, 89, 519-523.

Kinzig, A.P., Pacala, S.W. & Tilman, D. (eds.) (2001). *The Functional Consequences of Biodiversity*. Princeton University Press. Princeton, New Jersey.

Krebs, C.J. (1999). *Ecological Methodology*, 2nd ed. Addison-Welsey. Menlo Park, California.

Loreau, M., Naeem, S. & Inchausti, P. (2002). *Biodiversity and Ecosystem Functioning*. Oxford University Press. Oxford.

Pielou, E.D. (1969). *An Introduction to Mathematical Ecology*. Wiley-Interscience. New York.

Rosenzweig, M.L. (1999). Species diversity. In: J. McGlade (Ed.), *Advanced Ecological Theory* (pp. 249-281). Blackwell. Oxford.

Sanders, H.L. (1968). Marine benthic diversity: a comparative study. *American Naturalist*, 102, 243-282.

Simpson, E.H. (1949). Measurement of diversity. *Nature*, 163, 688.

Smith, B. & Wilson J.B. (1996). A consumer's guide to evenness indices. *Oikos*, 76, 70-82.

Wardle, D.A. (2002). *Communities and Ecosystems. Linking the Aboveground and Belowground Components*. Princeton University Press. Princeton, New Jersey.

CHAPTER 43

A PRIMER FOR STATISTICAL ANALYSIS

FELIX BÄRLOCHER

63B York Street, Dept. Biology, Mt. Allison University, Sackville, N.B., Canada E4L 1G7.

1. INTRODUCTION

Most scientific investigations begin with the collection of data. Summarizing and representing the data is generally labelled 'descriptive statistics'; conclusions, predictions or diagnoses based on these data fall under the domain of 'inferential statistics'. Inferences are never completely certain and are therefore expressed as probabilities. Consequently, to use statistical methods effectively, we need at least a basic understanding of the concepts of probability.

In every day life, we continuously make 'statistical' statements: we know, for example, that men tend to be taller than women, or that Scandinavians tend to have lighter skin than Egyptians. Such common-sense conclusions are generally reliable if the differences are large. Often, however, natural variability (environmental noise) is so great that it can mask the effect of factors that we investigate. Statistical evaluation is therefore essential, since our natural intuition can mislead us (Paulos 1995). For example, there is no scientifically justifiable doubt today that smoking poses a health risk. But we may still hear the argument that somebody knows a friend who smoked every day and lived a healthy life onto his 80s or 90s, and that therefore smoking may be harmless after all. We also tend to make unwarranted connections between a chance event and a particularly memorable success or failure: an athlete may have experienced a spectacular feat while wearing a particular sweater or pair of socks. Or, we may see a black cat and a few minutes later we have an accident. This tendency to interpret events in close temporal sequence as causally related can lead to superstitions, or prejudice – or, it may lead to new insights into actual mechanisms. Statistics can help us making rational decisions. It does not claim to reveal the truth. It has the more modest goal of increasing the probability that we correctly separate 'noise' from 'signal'. It helps us avoid both ignorance (being unaware of real connections between two variables) and superstition (accepting false connections between two variables).

M.A.S. Graça, F. Bärlocher & M.O. Gessner (eds.), Methods to Study Litter Decomposition:
A Practical Guide, 313 – 329.
© 2005 *Springer. Printed in The Netherlands.*

The way we evaluate chance and probabilities has been shaped by evolution (Pinker 2002). Attitudes that helped our ancestors survive and reproduce were favoured by natural selection. They were not necessarily those that infallibly separate signals from noise. To begin with, a complete evaluation of our environment would be time-consuming and exceed the capabilities of our central nervous system: "Our minds are adapted to a world that no longer exists, prone to misunderstandings, correctable only by arduous education" (Pinker 2002). Economists and psychologists refer to this shortcoming of our intellect politely as 'bounded rationality'. It plays an enormous role in many everyday choices and decisions. Investigations into how we perceive probabilities were pioneered by D. Kahnemann and O. Tversky (e.g., Kahnemann et al. 1982); Kahnemann was awarded the Nobel Prize in Economics for this work.

2. ROOTS OF STATISTICAL METHODS

The word "Statistik" was coined by G. Achenwall (Göttingen, Germany, 1719—1747). It is derived from "statista" (Italian for statesman), and refers to the knowledge that a statesman is supposed to have. Some early examples of statistical applications include population census, estimates of harvests in a country, taxes, etc.). Early statistical societies restricted themselves to the collection of data for economical and political purposes. They often deliberately refused to draw conclusions based on their data: the motto of the Statistical Society of London was "Aliis exterendum" – let others do the threshing, i.e., the extraction of conclusions (Gigerenzer et al. 1989, Bärlocher 1999).

An important breakthrough was made when Adolphe Quetelet (1796—1874) introduced the concept of the "average man", whose thoughts and deeds coincide with those of the entire society. He also recognized the importance of large numbers. Increasingly, the interpretation of collected data became important. The deliberate connection of measurements with probabilistic statements was initiated toward the end of the 19th century.

The impetus for probability theory came from games of chance. Its formal beginning is usually connected to an exchange of letters in 1654 between Blaise Pascal and Pierre de Fermat, discussing a gambling problem put to them by the Chevalier de Méré. The modern basis of probability was presented by Jakob Bernoulli (1654—1705) in Ars Conjectandi. Other important developments were the derivation of the normal distribution by de Moivre, and its further elaboration by Karl Friedrich Gauss. Thomas Bayes (1702—1763) introduced the important distinction between *a priori* and *a posteriori* probabilities. Bayesian Statistics, where *a priori* probability is often subjective, is well-established in economics and law. Its application to biology and other sciences is controversial.

Francis Galton (1822—1911) is considered the founder of eugenics and biometrics. Biometrics (or biometry) is defined as the application of mathematical techniques to organisms or life processes. Today, it is generally used more narrowly to describe the use of statistical methods in biological investigations. Galton developed the basis for regression and correlation. Another important technique, the χ^2 (chi square) test, was introduced by Karl Pearson (1857—1936).

The most influential theoretician of modern statistics is undoubtedly Sir Ronald A. Fisher (1890—1962). His work on analysis of variance, significance tests, experimental design, etc., continues to dominate the practice of data analysis (Zar 1996). His approach was modified and expanded by Jerzy Neyman (1894—1981) and Egon S. Pearson (1895—1980).

Statistics is often viewed as a monolithic, internally consistent structure of universally accepted concepts and laws. This is far from being the case (Gigerenzer et al. 1989). Deep-seated philosophical differences concerning the proper analysis and interpretation of data persist to this day, and no universally accepted approach seems to be in sight (Meehl 1978, Howson & Urbach 1993). What is represented as 'the' statistical method in textbooks has been called a 'hybrid theory', trying to reconcile the often contradictory approaches and interpretations by Fisher on the one side and Neyman/Pearson on the other side. Both differ from Bayesian statistics. A relatively new approach called model selection replaces traditional null hypothesis tests by simultaneously confronting several hypotheses by data. The enormous increase in computer power has allowed the manipulation of collected data and the production of 'synthetic' data, which may provide clues to their underlying structure (Monte Carlo techniques, Bootstrap, resampling and permutation methods; Efron & Tibshirani 1993, Manly 1997, Good 1994, 1999).

The development of powerful microcomputers and sophisticated statistical programs allows the application of very complex statistical models by naïve users. A taskforce of the American Psychological Association (APA, meeting on statistical inference, Newark, 14—15 December, 1996; http://www.apa.org/science/tfsi.html) saw this as problematic: the underlying assumptions are often ignored, little effort is made to determine whether the results are reasonable, and the precision of the analysis is often overestimated. The task force's recommendations include: making an attempt to verify the results by independent computation; more emphasis on simpler experimental designs; more emphasis on descriptive data analysis. This includes graphic representation (see Tukey 1977), calculation of averages with confidence intervals, and consideration of direction and size of effects.

3. FISHER'S APPROACH

3.1. Assuming Normal Distribution

How do we know that something is true? A naïve empiricist might reply that if we observe an event or a series of events often enough, it must be true. The Scottish philosopher David Hume (1711—1776) correctly argued that mere repetition of an event does not necessarily imply that it will occur in the future. An often used example concerns swans: Europeans are likely to encounter only white swans, and might conclude that all swans are white.

If repeated observations do not reliably reveal the truth, how do we decide which interpretation of nature is valid? The solution that has been accepted by most scientists (but see Howson & Urbach 1993, Berry 1996), and forms the basis of classical statistics, was suggested by Sir Karl Popper (1935). He agrees with Hume that our knowledge is always preliminary and based on assumptions or hypotheses. We can never verify these hypotheses. However, if a hypothesis does not represent the truth, it is vulnerable to being falsified. A useful hypothesis allows us to make predictions that are not obvious. We design an experiment to test these predictions; if they do not occur, we have falsified the hypothesis. For example, a European could propose the hypothesis that all swans are white. If he happens to visit New Zealand, he will sooner or later encounter a black swan, which falsifies his hypothesis. Or, as Thomas Huxley (1825—1895) put it: "The great tragedy of science is the slaying of a beautiful hypothesis by a nasty, ugly, little fact". Scientific research essentially is a weeding out of hypotheses that do not survive rigorous testing. Popper's reasoning was enormously influential. In economics, its basic philosophy has been expressed as follows: "The ultimate test of the validity of a theory is not conformity to the canons of formal logic, but the ability to deduce facts that have not yet been observed, that are capable of being contradicted by observation, and that subsequent observation does not contradict" (Friedman 1966). The same approach has been applied to natural selection: an organism, its organs and behaviour can be interpreted as 'hypotheses' concerning the nature of the environment. If they are inappropriate they will be 'rejected' by nature, i.e., the organism dies.

Biological hypotheses rarely allow yes or no predictions. Experiments more commonly produce continuous or discrete data, whose measurement cannot be accomplished without errors. Their true value must therefore be expressed in probabilistic terms. To take this into account, Fisher used the following approach:

- Formulate a null hypothesis (H_0). For example, we propose that two groups of animals on different diets have the same final body weight.
- Define a test statistic characterizing the difference between the two groups (the most obvious number to choose is simply the difference between the two averages; more commonly the t-value is used). Measure the actual value of this statistic.
- Assume that the weights of animals vary according to a defined probabilistic distribution (generally a normal distribution).
- Assuming that the two groups have in fact the same final weight (i.e., H_0 is correct), how likely is it that the test statistic will reach a value that is at least as extreme as the one actually measured (extreme is measured in terms of distance from the most probable value, which is the average)? This value, generally determined from the assumed data distribution, is called p.
- If p falls below a pre-established critical value α (frequently 0.05 or 0.01), we reject the null hypothesis. We label the two values as significantly different.

To repeat, p measures the probability that our test statistic (a number measuring a discrepancy between two or more groups) reaches a value at least as high as the one actually found **if** the null hypothesis is correct. It does not tell us anything about the probability that H_0 is correct or false. Because our measurements are always subject to random error, extreme values are possible and will occur. The value of α therefore also represents the probability that we incorrectly reject a null hypothesis that is in fact true (Table 43.1). According to Fisher, we can reject H_0, but we can never prove it to be correct.

Table 43.1. Statistical decision theory based on Neyman/Pearson (1933)

Decision	Null hypothesis (H_0)	
	Correct	False
Accept H_0	Correct decision	Type II Error (Ignorance)
Reject H_0	Type I Error (Superstition)	Correct decision

3.2. Assuming Data are not Normally Distributed: Permutation Tests and the Bootstrap

Most classical statistical tests assume normal distribution of the data (more accurately, errors or residuals that remain after a model has been fitted have to be normally distributed; in many cases, normal data imply normal errors and vice versa). If this is not the case, data can be transformed to make them approximately normal, or we can use non-parametric or distribution-free tests. The vast majority of these tests are variations of permutation or randomization tests (Edginton 1987, Westfall & Young 1993). Fisher again played a crucial role in developing this approach. The major difference to parametric tests is that we make fewer assumptions concerning the distribution of the data. Thus:

1. Formulate a null hypothesis (H_0). For example, we propose that two groups of animals on different diets have the same final body weight.
2. Define a test statistic characterizing the difference between the two groups (e.g., difference between the two averages). Calculate the actual value of this statistic.
3. Assuming that the two groups have in fact the same final weight (i.e., H_0 is correct), assignment of the measured values to the two diets should be random. We therefore systematically establish all permutations of the data. For each permutation, we determine the value of the test statistic.
4. How likely is it that the test statistic will reach a value that is at least as extreme as the one actually measured? This value, determined from the distribution of permutated data, is called p.
5. If this probability is below a pre-established critical value α (frequently 0.05 or 0.01), we reject the null hypothesis. We label the two values as significantly different.

Even with small data collections, an exhaustive listing and evaluating of all permutations can be extremely labour intensive. Before the advent of powerful computers, actual data were therefore first converted to ranks, which were then permutated. This generally results in a loss of statistical power (the ability to correctly reject a false H_0). With today's powerful microcomputers, actual data can be used. An extremely useful program, which allows reproducing almost all parametric and non-parametric tests, and the definition and evaluation of nonconventional test statistics, is Resampling Stats (www.statistics.com). A brief introduction is given in Section 6 of this Chapter.

Permutation tests are based on sampling without replacement, i.e., each collected value is used only once in a new 'pseudo sample' or 'resample'. Bootstrapping techniques use sampling with replacement. This means that collected values can occur more than once (Efron & Tibshirani 1993).

4. MODIFICATIONS BY NEYMAN AND PEARSON

Fisher's approach was expanded and modified by Neyman & Pearson (1933). In addition to H_0, at least one alternative hypothesis, H_A, is formulated. We proceed as follows:

1. Formulate a null hypothesis (H_0). For example, we propose that the final body weights of two groups of animals on different diets is identical.
2. Formulate at least one alternative hypothesis (H_A). For example, we propose that the final weights differ by at least 5 mg.
3. Define a test statistic characterizing the difference between the two groups. Measure the actual value of this statistic.
4. Generate the probability distribution of this test statistic assuming H_0 is correct and data are normally distributed.
5. Define a critical value of α. This again defines the probability of falsely rejecting H_0 (Table 43.1).
6. In addition, define a probability β of committing a Type II error. This is the probability of falsely accepting H_0. The term $(1-\beta)$ defines the probability of correctly accepting H_A; it is also called the power of the test (Cohen 1988).
7. Calculate p of the observed test statistic. If $p \leq \alpha$, reject H_0 and accept H_A. If $p \geq \alpha$, accept H_0 and reject H_A.

The Neyman-Pearson approach forces us to make a decision between two specified hypotheses. Depending on the costs or benefits of Type I and Type II errors, we will adjust the critical difference between H_0 and H_A, and α and β levels. For example, the commercial success of new drugs will depend on manufacturing costs and improved effectiveness (customers may be willing to pay double the price, if the drug is twice as effective, but not if the improvement is only 5%). In medical diagnosis, we must strike a balance between Type I errors (diagnosing a disease where none exists, false positive) and Type II errors (missing an existing disease, false negative). In law we have to balance the strength of the evidence (effect size) against the potential harm of letting the guilty walk free (Type II) or wrongfully convicting the innocent (Type I).

Some textbooks define H_A simply as the opposite of H_0, i.e., H_0: difference is 0, H_A, difference is not 0. This is not very useful, since it does not allow us to estimate β and $(1-\beta)$. In general terms, the power of a test increases with sample size, with effect size (difference between competing hypotheses), and decreases with the variability of data. Given sufficiently large samples, p will almost always fall below any specified α. A significance test by itself, without information on the effect size and its confidence limits, is therefore considered to be meaningless by many statisticians.

The free program G*power is useful to plan experiments (Erdfelder et al. 1996). For example, given α, β and effect size it estimates the number of samples necessary to detect a significant difference. Or, it estimates the power of an experiment, provided the other parameters are known. G*power can be downloaded at http://www.psycho.uni-duesseldorf.de/aap/projects/gpower/index.html.

5. BAYESIAN STATISTICS

As discussed above, classical hypothesis testing gives a p value that describes the probability of a value of the test statistic that is at least as extreme as the one found, provided H_0 is correct. But intuitively, we are more interested in the probability that H_0 is correct. It is a common mistake to assume that the two probabilities are identical. A simple example can illustrate this fallacy: scarlet fever is due to beta-hemolytic streptococcal bacteria. It usually causes a red rash, particularly on neck and chest. Suppose we observe a woman with a red rash. Applying statistical reasoning, we propose the null hypothesis that she does not have scarlet fever. If this is true, how likely is it she will have a red rash? Let us assume that a red rash occurs in 3 out of 100 of randomly selected women ($p = 0.03$). We might be tempted to reject H_0 and conclude that the woman has indeed been infected with scarlet fever. But the more relevant question surely is: if we observe a woman with a red rash, how likely is it that this rash is due to scarlet fever? To answer this, we would have to know all potential causes of a rash, which may include eczemas, pregnancy, allergies, measles, scarlet fever, and their relative contributions to rashes in the population. The contribution of scarlet fever to all rashes may be as low as 1%. The probability that the woman suffers from scarlet fever is therefore 0.01, and we should not reject H_0.

The Bayesian approach to statistics allows such direct probability statements. For example, we can estimate the probability that a new treatment is more effective than a control. We do this by modifying an initial estimate of this probability, which we base on prior experience or our intuition (for easy-to-follow introductions to this topic, see Berry 1996, Hilborn & Mangel 1997).

Bayes' formula (Bayes 1763) allows us to determine conditional probabilities: What is the probability of event A occurring given that event B has occurred, or, in mathematical notation $p(A|B)$? A common application is an estimate of the accuracy of medical diagnoses. Assume that a virus has infected 1 out of every 1000 people. This is the *a priori* probability: if we randomly chose a person, the probability that he or she is infected, is 1 in 1000, or 0.001.

A diagnostic test has been developed, which correctly identifies 99 out of 100 patients that have the infection. One out of 100 is incorrectly classified as non-infected. This is called a 'false negative'. The same test, applied to non-infected persons, correctly identifies 98 as healthy, but gives a wrong result in 2 cases out of 100; this person, though healthy, is diagnosed as being infected. This is a 'false positive'.

Now assume that we give a test to a randomly chosen individual in the population, and it turns out to be positive. What is the probability that the person is actually infected? This is the *a posteriori* probability.

Bayes (1763) developed the following formula:

$$p(A|B) = \frac{p(A) \cdot p(B|A)}{p(A) \cdot p(B|A) + p(notA) \cdot p(B|notA)} \tag{43.1}$$

where

$p(A)$ = probability of A

$p(A|B)$ = probability of A given that B has occurred

$p(B|A)$ = probability of B given that A has occurred

$p(notA)$ = probability that A has not occurred

$p(B|notA)$ = probability of B given that A has not occurred

In our example, $p(A|B)$ is the probability we are looking for: How likely is it that the person is infected, given the test was positive?

$p(A)$ stands for the probability of infections. In the population, it is 0.001.

$p(B|A)$ stands for the probability of a positive test given an infection. It is 0.99

$p(notA)$ is the probability of not being infected. It is $(1-0.001) = 0.999$

$p(B|notA)$ is the probability of a positive test in the absence of an infection = 0.02

We get the following result:

$$p(A|B) = \frac{0.001 \cdot 0.99}{0.001 \cdot 0.99 + 0.999 \cdot 0.02} = \frac{0.00099}{0.02097} = 0.0472$$

Surprisingly, the probability of infection is less than 5%! Nevertheless, this *a posteriori* probability is much higher than the *a priori* probability of 0.001. We have modified it by experience.

The same result can be found with the following approach, which may be more intuitive. Assume we are dealing with a population of 1 million and test every single individual. The results are summarized in Table 43.2, which also reveals the similarity of statistical tests to medical diagnosis. Similar examples could be constructed based on criminal trials (H_0 would correspond to innocence).

Table 43.2. Bayesian analysis applied to medical diagnosis. False negatives and false positives correspond to Type II and Type I errors, respectively.

Diagnosis	Infection		Total
	Absent	*Present*	
No infection	979,020	10	979,030
	True negatives	False negatives	
Infection	19,980	990	20,970
	False positives	True positives	
	999,000	1,000	1,000,000

Of all positive tests (20,970), 990 were true positives; therefore, $p(A|B) = 990/20970 = 0.0472$. We can apply this Bayesian perspective to interpreting statistical significance. Imagine that we are testing drugs for their ability to lower blood pressure. Based on the acceptable improvement and variability of the data, we have chosen a sample size that gives a power $(1-\beta)$ of 0.8, and we are willing to accept an α of 0.05. We run the test, and p is indeed < 0.05. How likely it is that rejection of H_0 is the correct decision? The short answer is: it depends on how much we already know about the drug. How likely do we think a positive result is going to be? Then we apply the same reasoning outlined above. The answer can be found at http://www.graphpad.com/articles/interpret/principles/bayes.htm. Motulsky (1995), Bärlocher (1999), and Brophy & Joseph (1995) provide more examples of how prior knowledge can modify p values determined from an experiment. For a more thorough introduction to Bayesian statistics, consult Berry (1996), Hilborn & Mangel (1997), and Howson & Urbach (1993). Potential applications to ecological studies have been reviewed by Ellison (2004).

6. MODEL SELECTION

Model selection replaces the traditional testing of a null hypothesis by confronting collected data with several competing models. The relative support of the data for each model is determined, allowing ranking and weighting of the models, and measuring the relative support for several competing hypotheses. Where similar levels of support are found, model averaging can provide robust parameter estimates and predictions. Model selection is based on likelihood theory and has been widely used and accepted in molecular systematics and mark-recapture analysis. Other potential applications in ecology and evolution are discussed by Hibbes & Mangel (1997) and Johnson & Omland (2004).

7. RESAMPLING STATS

7.1. Introduction

Resampling Stats is an extremely useful and powerful program to evaluate probabilities. It allows the reproduction of almost all classical tests, assuming normality, as well as non-parametric or distribution-free tests based on ranks or using actual data. It can also be used for bootstrapping or Monte Carlo techniques. More information is provided at http://statistics.com.

The basic idea is to generate or introduce numbers into a so-called vector, which is a collection of numbers and is given a name. Whenever we manipulate these numbers (e.g. by shuffling them, determining their average, etc.), the result is placed in a new vector, which has to be given a different name.

The Print command allows to check whether our commands are doing what we want them to do. Print commands should be deleted before executing the final program.

7.2. Some commands in Resampling Stats

- Data (1 2 3 4) A   Places the numbers 1, 2, 3 and 4 into vector A
- Shuffle A B   Shuffles numbers in A, places them in B
- Print B   Prints the result on screen when you select Execute

The same result can be found by writing:
- Shuffle (1 2 3 4) A
- Print A

To generate 1000 numbers of a normal distribution with an average of 2 and SD of 1, we write:
- Normal 1000 2 1 A

To prepare a histogram of the data in A, we write:
- Normal 1000 2 1 A
- Histogram A

To generate 10 numbers of a uniform distribution between 0 and 1, we write:
- Uniform 10 0 10 A

The next step calculates 2.5 and 97.5 percentiles of values in A (essentially values that enclose the central 95%; values outside this range correspond to the familiar $p = 0.05$).
- Normal 1000 2 1 A
- Histogram A
- Percentile A (2.5 97.5) B
- Print B

To take two random samples with replacement from 1, 3, 4, 5, 7, we write:

- Data (1 3 4 5 7) A   places data in vector A
- Sample 2 A B   takes two random samples from A and places them in B
- Print B   shows which two numbers were chosen

In addition, there are numerous statistical commands, such as Boxplot, Corr, Exp, Mode, Min, Regress, etc.

7.3. List of Commands in Resampling Stats

7.3.1. Basic Commands

ADD	Adds the elements of two vectors together
CONCAT	Combines two or more vectors into one long one
COPY	Records data or copies vectors; synonym for DATA and NUMBERS
COUNT	Determines frequency of a particular number or range of numbers
DATA	Enter data; synonym for COPY and NUMBERS
DEDUP	Eliminates duplicate values
DIVIDE	Divides the contents of one vector by another
END	Ends a loop, sends you back to a "REPEAT" statement
GENERATE	Produces the desired quantity of random integers within a defined range
HISTOGRAM	Produces a histogram of trial results
IF	Succeeding commands execute only when IF condition holds
MEAN	Calculates the mean of a vector
MULTIPLES	Determines the frequency of multiplicates of a number
MULTIPLY	Multiplies the elements in one vector by those in another
NUMBERS	Enter data or define a variable; synonym for DATA and COPY
PERCENTILE	Calculates the xth percentile of a frequency distribution
PRINT	Specifies output to be shown on screen
RANDOM	Produces the desired quantity of random integers within the desired range
REMARK	Allows user to insert a remark. Precede remarks with an apostrophe.
REPEAT	Allows user to repeat a simulation up to 15000 times
RUNS	Calculates the number of runs of a given length
SAMPLE	Samples with replacement
SCORE	Allows user to keep track of the result of each simulation
SHUFFLE	Randomizes the elements in a vector
SORT	Takes a specified number of elements from a vector and creates a new vector
URN	Creates a vector of specified quantities of specified numbers
WEED	Removes specified numbers or a specified range

6.3.2. Additional Mathematical and Statistical Commands

ABS	Finds absolute value of each element in a vector
BOXPLOT	Produces a Box plot
CORR	Calculates a correlation coefficient
EXP	Raises Euler's number, *e*, to the power of each vector element
EXPONENTIAL	Generates numbers from an exponential distribution
LET	Allows arithmetic expressions of the form LET x=a+b
LOGNORMAL	Generates numbers randomly from a lognormal distribution
MAX	Identifies the maximum value in a vector
MEDIAN	Calculates the median of a vector
MIN	Identifies the minimum value in a vector
MODE	Identifies the most frequent value in a vector
NORMAL	Generates numbers randomly from a normal distribution
PARETO	Generates numbers randomly from a pareto distribution
POISSON	Generates numbers randomly from a Poisson distribution
POWER	Raises each element in the first of two vectors to the power of the corresponding element in the second
REGRESS	Runs a multiple linear regression
RANKS	Computes the ranks of elements in a vector
ROUND	Rounds each element to an integer
SQRT	Determines the square root of each element in a vector
SQUARE	Squares each element in a vector
STDEV	Calculates the standard deviation of the numbers in a vector
SUMABSDEV	Sums the absolute deviations of one vector from another
SUMSQRDEV	Sums the squared deviations of one vector from another
TIMEPLOT	Plots a vector sequentially along the x-axis of a biplot
UNIFORM	Produces random values from a continuous uniform distribution
VARIANCE	Finds the variance of the elements in a vector
WEIBULL	Generates numbers randomly from a Weibull distribution

7.3.3. Additional Resampling and Housekeeping Commands

CLEAR	Erases contents of a vector
FUZZ	Sets precision for comparisons
MAXSIZE	Sets maximum vector size
PAUSE	Causes program execution to pause
PROGINFO	Provides information on memory use
READ	Imports data from an ASCII file
RECODE	Changes certain elements of a vector to specified value
SEED	Sets the random number generator seed
SET	To fill a vector with one value
SIZE	Determines the size of a vector
WHILE	Conditional repeat
WRITE	Exports data to an ASCII file

7.4. Three simple examples

7.4.1. Bootstrap Estimates of Confidence Intervals

Assume we wish to estimate the average height of a population of school children. We take a random sample of 15, and find the following values: 152, 140, 148, 134, 131, 156, 162, 150, 138, 153, 145, 153, 167, 143 and 130. The traditional method to estimate 95% confidence limits (CL) assumes normally distributed values; the mean and variance are estimated from the data. We get an average of 146.8, and the following confidence limits: 140.7 and 152.9.

With the bootstrap approach, we assume that the collected data are a true representation of the entire population. To reproduce this population, we simply multiply each measured value with a huge number (to approximate replacement); we then take a random sample from this pseudo-population. We determine the average of this bootstrap sample. We do this many times, keeping track of all averages. At the end, we determine the values that enclose the central 95%; these represent bootstrapped confidence limits. Instead of multiplying the collected data with a large number, we can sample with replacement. Thus:

```
MAXSIZE C 100000
REPEAT 100000
    Sample 15 (152 140 148 134 131 156 162 150 138 153 145 153 167 143 130) A
    Mean A B
    SCORE B C
END
Mean C Aver
Print Aver
Percentile C (2.5 97.5) CL
Print CL
```

A sample run gave an average of 146.8, and confidence limits of 141.3 and 152.3, remarkably close to the theoretical values. One important difference is that bootstrapped values can never go beyond the values that have actually been measured; the lowest possible average of any pseudo-sample would therefore be 130. No such restriction exists when we assume that data are normally distributed; theoretically, some children would be 1 cm, others 500 cm, tall.

7.4.2. Comparison of Two Groups

Assume we test the effect of a fertilizer on plant growth. We have 10 replicates each; the first column gives control data (no fertilizer), and the second column gives data with fertilizer:

44	55
56	47
58	63
34	62
49	49
61	63
56	73
43	68
53	59
49	48

With an unpaired t-test, we find a p value of 0.04 ($t = 2.205$, 18 degrees of freedom, two-tailed test). To test our null hypothesis with Resampling Stats, we combine all 20 data (H_0 postulates that they belong to the same population). We randomly subdivide the 20 data into two groups of 10, and determine the difference between the two group averages. We do this many times. This gives us the distribution of all possible differences. The original difference of the two groups was 8.4. How likely is it that by random redistribution of the original 20 values among two groups, we find a difference that is at least as extreme as 8.4 (if we are looking at a two-tailed test, this means ≥ 8.4 or ≤ -8.4)?

```
MAXSIZE Difs 10000
REPEAT 10000
      Shuffle (44 55 56 47 58 63 34 62 49 49 61 63 56 73 43 68 53 59 49 48) A
      Take A 1, 10 B
      Mean B C
      Take A 11, 20 D
      Mean D E
      Subtract C E Dif
      Score Dif Difs
END
Count Difs >= 8.4 high
Count Difs <= –8.4 low
Add high low Tot
Print Tot
```

A sample run gave a value of 414, i.e., 414 out of 10000 runs reached or exceeded the difference of the original data. This corresponds to a p value of 0.04 (414/10000), we would again reject H_0.

7.4.3. Analysis of Feeding Preferences
We wish to compare consumption of 3 food items using the experimental design presented in Chapter 41. Table 43.3 lists the daily consumption values of items A, B and C in 10 replicates.

The grand average is $(3.46+1.61+0.31)/3 = 1.79$. As test statistic S, we choose the sum of the squared deviations of the three measured consumption rates from the grand mean, i.e. $(3.46 - 1.79)^2 + (1.61 - 1.79)^2 + (0.31 - 1.79)^2$. For the original data, this gives a value of 5.033. Next, we estimate the probability of finding a value

Table 43.3. Consumption of food items A, B and C in 10 replicate containers

Replicate	A	B	C
1	3.57	2.35	0.48
2	0.35	1.87	0.40
3	3.14	0.31	0.67
4	7.17	2.29	0.28
5	3.24	3.46	0.35
6	3.07	1.11	0.17
7	5.69	0.55	0.03
8	4.45	0.85	0.34
9	1.48	1.47	0.12
10	2.48	1.85	0.22
Average	3.46	1.61	0.31

of S at least as extreme as 5.033 (≥ 5.033) if the assignment of the measured consumption rates to the three food items in each container were random, i.e., if there were no consistent preferences for one food item over the other. For each shuffled data set, we calculate the value of S. We do this 10000 times, which gives a reasonable approximation of the distribution of all possible S values. What proportion of this distribution has a value of ≥ 5.033? This proportion is equivalent to the classical definition of the p value. In the current example, this happens approximately 1—4 times in 10000 trials, therefore $p \approx 0.0002$. With Resampling Stats, the test can be run as follows:

```
MAXSIZE SSS 10000 'adjusts size of vector
REPEAT 10000 'random arrangement of data is run 10000 times
SHUFFLE (3.57 2.35 0.48) R1 'introduce and shuffle data from first replicate
SHUFFLE (0.35 1.87 0.40) R2 'introduce and shuffle data from second replicate
SHUFFLE (3.14 0.31 0.67) R3
SHUFFLE (7.17 2.29 0.28) R4
SHUFFLE (3.24 3.46 0.35) R5
SHUFFLE (3.07 1.11 0.17) R6
SHUFFLE (5.69 0.55 0.03) R7
SHUFFLE (4.45 0.85 0.34) R8
SHUFFLE (1.48 1.47 0.12) R9
SHUFFLE (2.48 1.85 0.22) R10

TAKE R1 1 A1 'take first number from first replicate, put it in A1'
TAKE R2 1 A2 'take first number from second replicate, put it in A2'
TAKE R3 1 A3
TAKE R4 1 A4
```

```
TAKE R5 1 A5
TAKE R6 1 A6
TAKE R7 1 A7
TAKE R8 1 A8
TAKE R9 1 A9
TAKE R10 1 A10

CONCAT A1 A2 A3 A4 A5 A6 A7 A8 A9 A10 A 'put all first numbers in A'
MEAN A AVA 'calculate average consumption of A in10 containers'

TAKE R1 2 B1 'take second number from first replicate, put it in B1'
TAKE R2 2 B2 'take second number from second replicate, put it in B2'
TAKE R3 2 B3
TAKE R4 2 B4
TAKE R5 2 B5
TAKE R6 2 B6
TAKE R7 2 B7
TAKE R8 2 B8
TAKE R9 2 B9
TAKE R10 2 B10

CONCAT B1 B2 B3 B4 B5 B6 B7 B8 B9 B10 B 'put all second numbers in B'
MEAN B AVB 'calculate average consumption of B in 10 containers'

TAKE R1 3 C1 'take third number from first replicate, put it in C1'
TAKE R2 3 C2 'take third number from first replicate, put it in C2'
TAKE R3 3 C3
TAKE R4 3 C4
TAKE R5 3 C5
TAKE R6 3 C6
TAKE R7 3 C7
TAKE R8 3 C8
TAKE R9 3 C9
TAKE R10 3 C10
CONCAT C1 C2 C3 C4 C5 C6 C7 C8 C9 C10 C 'put all third numbers in C'
MEAN C AVC 'calculate average consumption of C in 10 containers'

CONCAT AVA AVB AVC AVS 'combines average consumption values'
SUBTRACT AVS 1.79 DEVS 'determines deviations from grand mean'
SQUARE DEVS SDEVS 'squares deviations'
SUM SDEVS SS 'sums squared deviations'

SCORE SS SSS 'store all values of test statistic S in SSS
END
COUNT SSS >= 5.033  RESULT 'counts all S values ≥ 5.033, puts it in RESULT
PRINT RESULT 'prints how often S ≥ 5.033; this number/10000 is equivalent to p.
```

8. REFERENCES

Bärlocher, F. (1999). *Biostatistik*. Thieme-Verlag, Stuttgart.

Bayes, T. (1763). An essay towards solving a problem in the doctrine of chances. *Philosophical Transactions of the Royal Society of London* 66, 423-437.

Berry, DA. (1996). *Statistics: A Bayesian Perspective*. Duxbury Press. Belmont, California.

Brophy, J.M. & Joseph, L. (1995). Placing trials in context using Bayesian analysis. *Journal of the American Medical Association*, 273, 871-875

Cohen, J. (1988). *Statistical Power Analysis for the Behavioral Sciences*. Lawrence Erlbaum. Hillsdale, NewJersey.

Edginton, E.S. (1987). *Randomization Tests*. Marcel Dekker. New York & Basel.

Efron, B. & Tibshirani, R. (1993). *Introduction to the Bootstrap*. Chapman & Hall. New York.

Ellison, A.M. (2004). Bayesian inference in ecology. *Ecology Letters* 7, 509-520.

Erdfelder, E., Faul, F. & Buchner, A. (1996). G Power: A general power analysis program. *Behaviour Reserch Methods Instruments and Computers*, 28, 1-11.

Friedman, M. (1966). *Essays in Positive Economics*. University of Chicago Press. Chicago.

Gigerenzer, G., Swijtink, Z., Porter, T., Daston, L., Beatty, J. & Krüger, L. (1989). *The Empire of Chance*. Cambridge University Press. Cambridge.

Good, P.I. (1994). *Permutation Tests: A Practical Guide to Resampling Methods for Testing Hypotheses*. Springer-Verlag. New York.

Good, P.I. (1999). *Resampling Methods*. Birkhäuser. Basel.

Hillborn, R. & Mangel, M. (1997). *The Ecological Detective*. Princeton University Press. Princeton.

Howson, C. & Urbach, P. (1993). *Scientific Reasoning: The Bayesian Approach*. Open Court Publishing. Peru, Illinois.

Johnson, J.B. & Omland, K.S. (2004). Model selection in ecology and evolution. *TRENDS in Ecology and Evolution* 19, 101-108.

Kahnemann, D., Slovic, A. & Tversky, A. (1982). *Judgment under Uncertainty: Heuristics and Biases*. Cambridge University Press. Cambridge.

Manly, B.F.J. (1997). *Randomization, Bootstrap and Monte Carlo Methods in Biology*. Chapman & Hall. London.

Meehl, P.E. (1978). Theoretical risks and tabular risks: Sir Karl, Sir Ronald, and the slow progress of soft psychology. *Journal of Consulting and Clinical Psychology*, 46, 806-834.

Motulsky, H. (1995). *Intuitive Biostatistics*. Oxford University Press. Oxford.

Neyman, J. & Pearson, E.S. (1933). On the problem of the most efficient tests of statistical hypotheses. *Philosophical Transactions of the Royal Society of London, Ser. A*, 231:289-337.

Paulos, J.A. (1995). *A Mathematician Reads the Newspaper*. Doubleday. New York.

Pinker, S. (2002). *The Blank Slate*. Viking-Penguin. New York.

Popper, K.R. (1935). *Logik der Forschung*. Julius Springer. Vienna.

Tukey, J.W. (1977). *Exploratory Data Analysis*. Addison-Wesley. Reading, Massachusetts.

Westfall, P.H. & Young, S.S. (1993). *Resampling-based Multiple Testing*. Wiley & Sons. New York.

Zar, J. H. (1996). *Biostatistical Analysis*. Prentice-Hall. Upper Saddle River, New Jersey.